SpringerBriefs in Environmental Science

More information about this series at http://www.springer.com/series/8868

Badri Narayanan Gopalakrishnan

Economic and Environmental Policy Issues in Indian Textile and Apparel Industries

Springer

Badri Narayanan Gopalakrishnan
School of Environmental and Forest Sciences
University of Washington-Seattle
Seattle, WA, USA

ISSN 2191-5547 ISSN 2191-5555 (electronic)
SpringerBriefs in Environmental Science
ISBN 978-3-319-62342-9 ISBN 978-3-319-62344-3 (eBook)
DOI 10.1007/978-3-319-62344-3

Library of Congress Control Number: 2017947904

Printed on acid-free paper

This Springer imprint is published by Springer Nature
The registered company is Springer International Publishing AG
The registered company address is: Gewerbestrasse 11, 6330 Cham, Switzerland

Dedicated to Sumathi, Anagha, Hari and my parents

Preface

In this century, India and China have risen as the major emerging economies of the world, apart from Brazil and Russia. Among these, India is poised to sustain as the fastest growing economy in the world according to several forecasts including that by the International Monetary Fund (IMF). Hence, studying the Indian economy is essential to understand the future of the global economy.

Although India is better known globally in recent times for its information technology and services sector, agriculture and manufacturing are still major parts of India's economy, more so than the services sector. The new Indian government, sworn-in in mid-2014, has vowed to rejuvenate the manufacturing sector through efforts such as the 'Make in India' campaign to promote investment and production in Indian manufacturing, and 'Mudra Bank' for small-scale investment.

Within the manufacturing industry, the textile and apparel sector is a major contributor to India's employment, GDP and export earnings. Various structural changes have occurred in this sector over the past two decades in India, at global and national policy levels. In this book, I discuss various economic and environmental policy issues in this significant sector in the Indian economy, which in turn plays a major role in the global economy.

This book is expected to benefit people from several walks of life, including industrialists, economists, environmental scientists, public policy enthusiasts and analysts, other social scientists and academicians, and students of several subjects in policy, business and economy. In short, anyone with some inclination towards policy studies in developing countries in general, and trade, industrial, environmental and economic policies in particular, may possibly find this book quite useful even if they do not have specific interest in the textile and apparel industry.

Having curated the usefulness and relevance of the topic of this book into a broad perspective, with the target audience also in mind, a synthesis of the whole book is provided. Chapter 1 discusses the historical developments in the textile and apparel industries and links them with the current structure of these industries. This chapter details the historical aspects, starting from ancient times until modern times, of the textile industry in India. The objective is to understand how the features of the industry today are related to the features of past. Specific observations that link the present with the past are included, considering both the policies and structure of the industry.

Chapter 2 provides an overview of the technological aspects of this sector, including its definition and scope. It also deals with an overview of the economic structure of Indian textile and apparel industries. Statistics on Indian textile and apparel industries are reviewed to make inferences on their overall economic structure.

Chapter 3 presents an economic and social overview of the Indian textile and apparel sector. It covers four broad areas: employment, the unorganized sector, trade liberalization and demand. Trade in textiles was opened up on two counts in India. Firstly, the external restrictions of export quotas (MFA or the Multi-Fiber Arrangement) were removed in 2005, in phases since 1995. Secondly, India has been reducing barriers on most of its imports since 1991. Both of these have had profound impacts on the Indian textile industry, as discussed in this chapter. I also discuss vulnerable sections in the textile industry, focusing on the unorganized or informal sector that is huge in the Indian textile industry. I analyze some facts and figures and compare the organized and unorganized sectors. This chapter also discusses the employment issues and policy aspects. As the second largest employer in India, the textile industry has huge implications for its industrial and labour policies. This chapter covers these aspects. I further focus on the demand-side factors. While a lot has been discussed about the supply-side aspects, demand for textiles is a major issue. There has been genuine concern among policymakers and academicians that the textile industry in India is constrained by domestic demand.

Chapter 4 elaborates on the environmental issues in this sector. While every chapter in this book has an environmental dimension, this chapter focuses exclusively on these issues through a review of literature and with some fresh analysis.

Seattle, WA, USA   Badri Narayanan Gopalakrishnan

Contents

Chapter 1
History of Indian Textile Industry

The cotton industry has always occupied a honoured place in Indian industrial system. India is the accredited birthplace of cotton manufactures. Certain passages in Rig Veda, Mahabharata, Ramayana, Puranas, etc. indicate the antiquity and flourishing state of the industry from the prehistoric times (Govil 1950). Ray (1986) explains that ancient Indians were using dyes such as indigo, lac, turmeric, madder, resins and red ochre. Varahamihira (500 AD), in his celebrated Brihat Samhita, refers to mordants such as alums and sulphates of iron to fix dyes on textile fabrics. During the Buddhist period, the Indian cotton fabrics were exported worldwide, according to the records of various travellers and ambassadors a few centuries ago. The industry was fortunately not much affected by the wars and rebellions during the Muslim invasion, as this was relatively short, and was patronized by the Mughal emperors, who were lovers of luxuries and fine articles. Daca, Masulipatnam, Madras, Gujarat, Ahmedabad and Banares emerged as prominent textile centres. Indian cotton fabrics were known for their fine quality.

Indian textile industry was so advanced as early as the sixteenth century that the techniques such as mordant fixing of dyes in cotton clothes were copied from Indian craftsmen on the Malabar coast in the 1600s (Chapman 1972). By 1712, East India Company was informing its agents that printing could be done in England at half the price charged for Indian goods and in better colours and patterns. The impact of imperialism on Indian textiles may be well-illustrated by the fact that India took only 0.5% of British cotton cloth exports in 1815, but this increased to 23% and 31% in 1850 and 1860, respectively.

The size and diversity of handloom weaving industry in India had attracted immense part of European trade in seventeenth and early eighteenth centuries. Roy (1996a) notes that despite the exports to Europe in the pre-1800 period, Indian weavers faced a severe competition from foreign producers since the mid-eighteenth century, because of which they could have adopted better strategies such as diversification of their products, lest avoid being substitutes to imports from England and suffer from a fall in demand for domestic products owing to fall in import prices due to mechanization.

© The Author(s) 2018
B.N. Gopalakrishnan, *Economic and Environmental Policy Issues in Indian Textile and Apparel Industries*, SpringerBriefs in Environmental Science, DOI 10.1007/978-3-319-62344-3_1

The four major cloth exporting regions in India before European trade were Punjab, Gujarat, the southeastern coast and Bengal, which were trading with Central Asia and the Middle East, Red Sea Ports, Southeast Asia and North of India, respectively. Their comparative advantages were resources such as quality of raw cotton in Bengal, water in most places, and specialization in serving distinct networks of long-distance trade (Chaudhuri 1996).

In all these regions, political insecurity and famines led to migration and deaths of many weavers, who were already gaining much littler than merchants who were taking care of the sales of their products. The production was mostly based on contracts between merchants and buyers, wherein non-delivery was included as a possibility of risk, and an advance was demanded as the weavers required working capital. Poor transportation system, lack of raw material supplies, as cotton harvests heavily depended on monsoon, and poor coordination between these parties led to the breaking of such contracts many a times by the late eighteenth century. There were also other complex interplays such as those between food-grain harvests and prices, cotton production, spinning, weaving and cloth prices. Hence, scale, quality control, standardization and risk measurement were major issues involved in this period.

In addition, there were many intermediaries between a merchant and a weaver and many other processes such as cotton farming, ginning and spinning, increasing the complexity involved. The situation changed gradually by the end of the eighteenth century, wherein artisans were directly employed by the merchants for regular wages. Many components of the traditionally and hereditarily imparted techniques such as dye fixing by mordants, specialty yarn spinning by hand and complex designs using simple looms were all contributions of India to the Western world in the area of textile science and technology.

During the Company Raj and even before it, from 1616 to 1757, the government policies on raw silk and silk weaving, pertaining to marketability, technology, organization, employment and capital structure of the industry, had some merits owing to reforms and many demerits (Ray 2005). Before European entry, Indian textile trade was erratic, seasonal and highly unstable, and hence this occupation was in addition to cultivation, as it offered low prices.

In Tamil Country, there was a hierarchy among the people involved in weaving: weavers with or without looms, caste heads, village leaders, weaver-trader master weavers and merchants. As this resulted in huge costs for the British, they appointed paid middlemen to have a closer control over producers. However, they faced stiff opposition from the local weavers of all levels, leading to migration away from the British territories to Pondicherry, though there was a fall in relative cloth prices. Agents were replaced by salaried procuring employees in Bengal in the late eighteenth century. Indian weavers retained their freedom in choosing the raw materials, tools and techniques, but prices, production and stocks were not controlled much by them (Arasaratnam 1996).

Facing a slump in Europe, Madras weavers began to supply cloth to Southeast Asian markets and exported idle weavers to those regions, unlike Bengal. They resorted to produce coarser clothes in response to the machine-made English cloth until 1850s. Factors such as general economic decline reflected in falling prices, redistributions with the decline of the courts and the export crisis could have hurt fine weaving. A good deal of product differentiation as well as focus on local and low-quality demand was happening to survive stiff competition from the local producers as well as imports in the late nineteenth century. However, the production of standard white cloth was widely affected because of competition from imports (Specker 1996).

Madras weavers seemed to shift towards manufacturing fine cotton cloth with golden thread borders, viscose garments, exports of items such as Madras handkerchiefs and lungies, as well as coarse cotton, depending on the location, as coarse weaving was widespread unlike fine weaving (Yanagisawa 1996). Diversifications were in many forms, such as Brahmanic clothing, thread-dyeing and innovations in fibres such as art-silk and mercerized yarns. Further, different small regions tended to specialize in different product categories. Hence, the south Indian handloom weavers grew in number despite a severe competition from local and imported mill-made cloth, in the early twentieth century. The social changes such as empowerment of scheduled castes in choosing their attire added to the demand for clothing in the mid-twentieth century. The condition of weavers in North Coromandel was similar (Swarnalatha 2005).

Though the preconditions to merchant capitalism existed even before, the entry of European companies expanded the cotton textile demand, to an unprecedentedly high level, so much so that the merchants were financially ruined owing to the technological constraints on the total textile production. However, the European companies had facilitated the creation of Port enclaves at Madras and Cuddalore, which not only ensured safe and hence prosperous places for the merchant capitalists in a (then) perennially warring country but also aided them to reinforce their high social position (Mukund 1999).

Bengali weavers were alienated in the process of conflict between the revenue and commercial interests of the British and the local zamindars by the early nineteenth century (Hossain 1988). Company's attempt to enforce low wages and maintain strict controls over weavers led to the decline of Bengali weaving industry in the early nineteenth century. Though this reduced the costs for the Company, it also led to large-scale attempts by weavers to abscond or default, resulting in sharp drops in quality and quantity of cloth supplied (Hossain 1988).

Starting from 1757, Mitra (1978) describes the decline in the traditional cotton industry in Bengal till 1833, when the East India Company lost its trading status because of demand constraint, competition from much cheaper machine-made clothes and no tariff protection. However, the control that British power had over Bengal weavers was unprecedented, and this led to many structural changes. The low prices offered and very specific commodities demanded by the British, in comparison with other European companies, made the lives of the cotton weavers miserable, especially after the 1830s.

Roberts (1996) explains that the French used the indigo-dyed long cloth called guinee produced in Pondicherry, to pay for the gum from West Africa, which was required as mordant to manufacture guinee as well, which had become a unit of exchange by the mid-eighteenth century. This shows the importance of textiles in colony-to-colony trade, early industrialisation in Pondicherry textiles and probably the first instance of export-oriented industrialisation strategy in India.

Guha (1996) notes that owing to the competition from the machine-made textiles in England, weavers in Bengal and the eastern coast had to switch from exports to domestic markets, while Central Provinces always supplied local market. The cotton famine in the 1860s, which drove up the cotton prices immensely, and the establishment of railways in 1860s, which provided alternative means of livelihood to weavers, led to really severe competition from imports, leading to a crisis in central Indian weaving with large-scale emigration and fall in earnings. Hand spinning disappeared by the 1920s, though hand-spun yarn was widely considered to be much more superior to machine-spun yarn. However, the weaving sector displayed an increase in notional employment between 1867 and 1926 and increased productivity thereafter. Since the 1930s, power looms started entering the scene and enhanced the use of mill-made yarn. By the 1940s and 1950s, the handlooms and handloom weavers in Central India declined in number by half, as compared with the early twentieth century, in Central India.

Gillion (1968) explains that Ahmedabad emerged as a major textile centre in India, despite its limitations such as dry climate, long distance from Bombay port, lack of availability of fuels and long-staple cotton, high railway tariffs, and very little investment, attention and promotion by British towards Ahmedabad. The development was primarily attributed to the common sense, practical mindedness, perseverance and perfection (at all stages) of the vaniyas in Ahmedabad. Broach in Gujarat was known for its excellent fine-quality textiles, bleaching and dyeing, since sixteenth century (Gokhale 1979). However, the Gujarati weavers were unable to comply with the requirements of British East India Company, while the Dutch were ready to collect whatever was available in the market.

Pearson (1976) hypothesizes that despite the fact that the rulers of Gujarat were powerful enough to end the Portuguese control of Asian maritime trade, they refrained from doing so, possibly because they thought that Portuguese activities did not threaten their interests. This gives a new way of looking at the impacts of colonialism on textile trade in India.

Sahai (2005) elucidates impacts of the state and politics on the wajabi system, which is a set of legitimate practices, on the artisans in the eighteenth century Jaipur. Artisans and the state agencies deployed this system to ensure the observation of normative conduct in intercaste relations and interactions. The wajabi concept was redefined and contested among artisans and the state. Though the talented and skilled artisans were rewarded by the kings in Marwar, there were huge taxes in Jodhpur, in addition to non-payment of wages and other means of harassment. This led to the emigration, relocation and even collective resistance by the artisans. Weavers' livelihoods in Banares have been under threat since the 1880s, but silk weavers have increased in number, due to many reasons including the uniqueness of these products and place (Kumar 1995).

Weavers tended to diversify towards more skill-intensive weaving and market shares of handloom cloth did not decline in the 1930s. The dynamism of and differentiations within the Indian weavers are evident from the rise of small-scale power looms in this period. Average prices of handloom cotton cloth and the amount of cotton used in blends consistently increase in the 1930s. Demand for cotton and non-cotton yarns by handloom and power loom sectors increased in the 1930s, showing that expansion of power looms could not be thought of having affected handlooms adversely. Similar trends are noted for silk and man-made fibres as well, implying that there was a structural change happening in the textile sector in India in this period. However, the production of hand-spun yarns declined sharply in the 1930s (Roy 1996b).

While total quantity and value of handloom products had risen in the 1930s, this was probably because of an expansion of non-cotton sector at the expense of cotton sector, as seen by the increase in quantity and value in the former and decrease in those in the latter. Power loom output had risen sharply, but with a steep fall in prices as well, while the mill cloth output grew rapidly in this decade. The structural changes that took place in textile sector had implications for various other allied occupations. For example, Haynes (1996) addresses the rules of exchange and relations of production in the case of an allied craft, namely jari.

Import of machine-made cotton yarn was continually increasing in South India till the late nineteenth century due to the fact that there were not many factories. However, hand-spun yarn was much better in comparison with the imported machine-spun yarn, because of low costs of production and unique benefits of Indian cotton and weather. Nevertheless, imported machine-spun yarns were used because they were essential for trade, though the hand-spun yarns were still extensively used to serve the domestic market. Hence, towards the end of the nineteenth century, the Indian textile sector remained a major employer, despite all its crises and fundamental changes in its structure, with complex effects on the weavers.

The first cotton mill was started in 1838 near Calcutta, followed by Bombay Spinning and Weaving Mill in 1851, which started operating since 1856. Boom in money lending and transport encouraged mill sector in places such as Ahmedabad soon after. The stringent labour legislations in Indian industries have their origins in the Indian factories act of 1881 that protected child labour that of 1891 that gave protection to female labour and that of 1911 that protected all types of labour, in order to protect the UK industry from the fast-growing Indian textile mill sector, in addition to introducing zero-customs for UK's imports to India. From 1896 to 1926, there was a gradual change in fiscal regime as custom duties were imposed on UK's exports as well, and the corresponding countervailing excise duties, levied on Indian cotton goods, were removed later. By 1900, there were 193 mills in India, half of which were in Bombay. The rise of Japanese textile industry led to the loss of Chinese market for Bombay's textile mills. Despite this and other troubles such as plague, famines and currency troubles till 1905, a short boom after 1905, till the first world war, led to the existence and functioning of 271 mills, with 68 lakhs spindles and one lakh four thousand looms (Govil 1950).

Famine Commission of 1880 and 1901 emphasized the need to industrialize India, to combat famines. As a response to this, the Imperial Department of Commerce and Industries started in 1905, coupled with the active Swadeshi movement, caused industrial regeneration (Majumdar et al. 1963). However, owing to the pressure from the British Government, the government intervention in the industrial sector was withdrawn, and laissez-faire was brought back in 1910. It was only after the First World War that the British realized that Industrial development of India was more in their own interests than those of Indians, as explained in the next paragraph, resulting in the setting up of Industrial Commission in 1916 that led to reforms in fiscal aspects and labour laws.

During the First World War, there were supply constraints as cheap machinery and dyestuff from Germany, Austria and Hungary ceased to be available, and hence the industry was stagnant. Further, the war bonus to the workers led to increase in costs, without great increases in prices. The initial years of the post-war period were prosperous for Indian textiles despite monsoon failure and influenza epidemic in 1918, due to the worldwide boom, low cotton prices, high cloth prices and a stronger Rupee, facilitating necessary imports. However, wages did not rise as much as profits did. Expansion of weaving, coupled with an existence of backward linkages, meant expansion of spinning. The contribution of mill sector to the total domestic textile requirements rose from 9% in 1899 to 42% in 1922, while that of the imports fell from 64% to 26% and that of the handlooms increased from 27% to 32%.

Due to the fact that Japan was a major exporter to India and Yen appreciated from 1921 to 1925, the domestic textile sector was adversely affected during these years, due to cheaper imports that could easily substitute the domestic textiles, which were already in a pathetic position given the fact that the fluctuations in cotton and cloth prices were like pendulum oscillations, i.e. higher cotton prices accompanied with lower cloth prices leading to heavily shrunk margins. There were other factors related to Japan, such as lack of labour regulations, subsidies, better organization and utilization that created severe competition compared to Indian textiles.

Based on some recommendations of the tariff board, Indian textiles were protected from Japanese imports by introducing higher customs duties, though the same was not done for British imports, justified by an argument that they were not the major threats. Various changes in the trade regime in terms of increased protection by means of increasing the customs duties of both British and non-British textiles as well as fixing quotas against Japanese imports, which is against Indo-Japanese trade convention of 1904, prompted Japan to boycott Indian cotton imports. In 1934, things were set right, as India reduced the duties and Japan called off the boycott. Indo-Japanese Trade Agreement of 1937 offered the most favoured nation status to Japan and retained the existing levels of quotas and tariffs, mentioning nothing about non-cotton textiles.

Several steps to pacify Indian textile producers as regards the preferential treatment given to British goods include Lees-Mody pact, Indian Tariff (Textile Protection) Amendment of 1934 and Indo-British Trade Agreement of 1935. However, none of these increased the tariffs on the British imports, despite strong opposition from the Legislative assembly and even a Tariff Board, till an agreement signed in 1939, which allowed for increased tariff if the imports are higher than a threshold.

During the Second World War, manufacturers and dealers hoarded textile goods, anticipating positive demand shocks after the war as it happened in 1919, only to drive the prices up, a phenomenon that could partly be attributed to fall in imports from Japan as well. Further, world demand for textiles, in addition to military purchases worldwide, also grew so much so that UK and Japan could not supply them, adding to further enhancement of demand for Indian exports, thereby reducing the amount available for domestic consumption. Handloom weavers lost employment because there was little yarn supply and net of exports.

Cotton textile advisory panel was created in 1941 to come up with solutions to enhance supply of textiles in India. This resulted in control of prices and procurement of textiles to a great extent. The concept of 'standard cloth' that involved purchasing fixed amount of clothes at a price that can be change only once in a quarter was introduced in 1943 and aided very much in stabilizing the prices and supply of cotton piece goods. Various committees and boards including the Textile Control Board and officers including the Textile Commissioner were started in the early 1940s to control the industry.

After the war, various controls were imposed, such as Textile Industry (control of production) Order of 1945 that warranted the mills to produce 'utility cloth' and reduce product diversification. In 1947, standardization scheme was introduced, further restricting the production. After independence, in 1948, all controls in prices, supplies, production and standards, including the Textile Control Board itself, were abolished, leading to rapid price hikes, forcing the government to revert back to the price controls. Tariffs that protected the mill sector were removed in 1948, in the interests of handloom weavers.

Handloom sector is perhaps the most relevant residual of the glorious past of Indian textiles. This sector is now mostly unorganized; lacks raw materials, marketing facilities, training and financial assistance; and suffers from greedy middlemen. This feature stems from the historical structure of the industry discussed in the previous section. Weaving is done either by independent weavers who decide on raw materials as well as end products, master weavers who undertake job-work by supplying raw materials to the weavers or by giving advanced cash credit to the weavers to manufacture specific quantity and quality of products. They suffer from lack of proper markets as they are generally localized, depend on middlemen and cannot afford credit payments in the case of distant markets. Cooperative societies are helpful in many ways such as providing rebates during times of low demand, but they do not function as efficiently as they could have, because of conflicting interests among their members.

To ensure proper supply of raw materials, the government started yarn depots in 1973, but they failed as yarns of required specifications and numbers were seldom available. Further, other raw materials such as dye and jari were not supplied under this scheme. Integrated Handloom Development Programme was not very successful owing to various bureaucratic hurdles. In addition to general poverty, illiteracy and other problems outlined above, the handloom weavers suffer from other setbacks such as lack of advertising, quality and finance (Niranjana and Vinayan 2001). Strikingly, most of these features existed even centuries ago, when the British arrived in India.

Handlooms have survived in India despite many modern developments and competition from mill and power loom sector, probably because of its dynamic nature and sustained demand in certain social spheres. They have been constituting 25% of total textile production for the past few decades, which is commendable given the fact that the textile industry itself is expanding, though this may be mainly because of the protection offered to the handlooms by the state (Mukund and Sundari 2001).

However, power loom sector remains a big competitor as it makes use of the concessions given to the handlooms, is protected by mill-owners as well and is primarily involved in synthetic fabrics, which form an increasing share of consumer demand in India, in addition to the obvious advantage of superior productivity at lower costs of labour. Further, most statistics about handlooms may include power looms as well, because the delineation is blurred, and states may be doing this to overstate the handloom performance in order to derive benefits of central government's schemes for handlooms. Government has taken steps to enhance the organization, modernization, marketing assistance and welfare of the handlooms.

Most textile policies and reports, such as the one by Kanungo Committee, were based on a premise that handlooms are inefficient means of weaving and require protection and/or modernization lest these weavers lose their means of livelihood. While the textile policy in the 1950s was to provide clothing to masses in adequate quantities by promoting handlooms, the 1978 textile policy recognized the need to enhance the per capita cloth consumption. The textile policy of 1985 noted the over-performance of the power loom sector and brought it under regulatory framework, though the enforcement of this provision is questionable till date. This policy statement argued for the promotion of finer quality and high value-added textiles that would increase weavers' earnings, contradicting its own action of reserving Janata cloth production for handlooms, which enhanced employment but at low wages.

New small enterprise policy of 1991, the report of High-Powered Committee on Handlooms of 1996 and the report of the working group on the handloom sector were all in the same vein in that they all stressed on the low economic and high cultural, social and ethical value of handlooms and that these weavers should focus on high-end export market, which does not seem to be recommendable, as learnt from the past and as handlooms are predominantly meant for domestic market now.

There used to be a regulation for Indian spinning mills, which required them to produce a fixed fraction of yarn in hank form, so that the handlooms do not suffer from shortage of yarn. Known as hank yarn obligation, this regulation came into place in 1974; it was fixed at 50% of the total marketable yarn in 1986, though it was brought down to 40% and then 20% in 2003. Despite hank yarn obligation, partly because of the power loom sector, the yarn supply to handlooms was so big of an issue that many weavers committed suicides in Andhra Pradesh, due to high yarn prices that washed out their livelihoods. This is also due to the stress on export-oriented spinning, which diverted a major part of yarn to export market.

While cooperatives should be provided with adequate supplies for yarn and dyes, the credit supply need not be subsidized as long as it is transparent and involves less red tapism. Further, prejudice against master weavers requires to be avoided as they have played a vital role in finding proper markets for handlooms.

Silk industry, which played a major role in rural Bengal till the late nineteenth century, started declining since then, not only because of competition from machine-made silk garments that restricted silk producers to sericulture and reeling, rather than the entire gamut of processes involved in producing silk garments, but also owing to various policy drawbacks of colonial as well as democratic governments. The major drawback was a complete failure, on the part of the government, to understand the industry from the viewpoint of the actual silk producers. Despite this, the remnants of silk industry in this region serve as much of livelihood as agriculture does, if not more, for the poor farmers. This is because of the demand for raw silk and silk yarn, abysmally low wages that reduced the costs, its high luxury value and its prominent dependence on agriculture (Schendel 1995).

The state consistently assumed the households in the silk industry to be self-sufficient rural capitalist enterprises and hence never provided things other than infusions of credit for the initial capital, know-how and technology. Exploitative marketing conditions, government's ignorance of impacts of other economic developments on silk sector, gender bias and lack of encouragement of participation of silk producers in decision-making.

India is the second largest producer of raw silk in the world, though it consumes most of its production. However, Indian silk industry severely suffers from very low yield in terms of mass per hectare of area (35 Kg/Ha in India compared to 100–120 Kg/Ha in Japan). Imported high-quality raw silk is illegally sold in the domestic market mostly to power looms. A high domestic demand, combined with declining supply of high-quality raw silk from countries like China, which increasingly focus on value addition, is now creating pressure to improve the quality of silk in India. States such as Tamil Nadu, Karnataka and Uttar Pradesh are leading in Silk production in India (Naik and Babu 1993).

From 1950 to 1980, the number of spinning mills increased fourfold, while weaving and composite mills suffered stagnation, due to the government-backed freezing of mill output and expansion of power loom sector that led to increased yarn demand. Supply constraint has dominated in this period, as cotton availability has fallen over years. Maharashtra, Gujarat and Tamil Nadu are the major states involved in cotton textile production, of which Tamil Nadu emerged after the 1960s. Clothes of coarser counts are produced in the mills, while handlooms and power looms produce finer ones. The secular decline in per capita consumption of cotton cloth, despite a rise in per capita real income, is due to its substitution by man-made fibres and blends, rise in prices of food and cloth as well as adverse income distribution. Price trends are increasing for cotton mill cloth, while they are decreasing or stagnant for synthetics and blends. High growth rates of food and cloth prices, coupled with decline in agricultural income, seem to have reduced cloth consumption in the sixties and seventies.

Declining cloth exports and growing yarn exports have preceded the emergence of garment exports since 1970. India's major markets for fabrics are the UK, USA, Canada, Australia, New Zealand and a few African countries. India's share in the global textile market declined till the 1980s, not only because of MFA quotas but also because of various other policy handicaps and productivity gaps. Emergence of sick mills is because of continuing inefficiency and poor performance of several weak and marginal units. This is due to inadequate depreciation provision, continued neglect of proper maintenance, repairs and replacements, mismanagement, lack of modernization, bankruptcy, focus on short-term gains and bulk production of controlled cloth. Power looms have reaped the benefits that were meant for hand-looms (Sastry 1984).

After World War II, there were many bilateral trade agreements among countries, till 1961, when a regulatory framework named Short-Term Agreement, was signed by Global Agreement on Trade and Tariff (GATT) member countries. This was replaced by Long-Term Agreement since 1962, which imposed controls on exports of cotton textiles and exports to the developed countries from the developing ones. Multi-Fiber Arrangement (MFA) came into force in 1974 to exercise controls and restrictions over imports of non-cotton textiles as well.

The first stage of MFA, which was till 1977, promised increase in export earnings for developing countries, with due considerations of market disruption that might occur owing to excessive imports to the developed countries. In such cases, the developed countries were empowered to restrain the levels of exports, based on the past exports, allowing for some positive growth rates as well. These could be done by bilateral consultations and these did apply for handlooms.

The second stage of MFA was from 1978 to 1981 and was more restrictive than the first one, as it allowed reasonable but temporary departures from the general terms of MFA. As the departures were mostly restrictions and were of continuing nature, this was detrimental to the export performance of the developing countries.

The third stage of MFA, which was from 1982 to 1986, was supposed to be less restrictive as it gave more provisions to the developing countries to be compensated for the safeguard measures. Textiles and apparel sectors were treated as two distinct sectors and quotas were worked out accordingly. However, this worsened the situation as regards Indian textile and apparel exports, as most bilateral agreements signed consisted of rigid features on category ceilings, growth rates, carry over, carry forward and swing provisions.

During its last stage, there were increasing resentments across the world against MFA, since it had allowed the developed countries to export among themselves without restrictions and to safeguard against all low-price exports. Even the consumers of developed countries were at loss, as they had to pay unnecessarily high prices due to these quotas. Thus, phasing out of MFA quotas was scheduled from 1995 till 2005, based on the Agreement on Textiles and Clothing. The increase in growth rates of all the categories, as agreed, was 16% from 1995 to 1998, 25% from 1999 to 2002 and 27% from 2003 to 2005. The importing countries could postpone the phasing out of certain sensitive categories, selected at random by them. Phasing out of MFA quotas is expected to increase the exports of textiles and apparel from developing countries such as India. Low domestic demand, high cotton prices, fiscal policies skewed against synthetics, quality issues and infrastructure bottlenecks are the problems faced by the industry today (Gokhale and Katti 1995).

There have been various committees, which have submitted reports on various textile policies of the government. For example, GoI (1990) notes that fluctuations in cotton prices, quality of cotton, prices of synthetics, decline of handlooms, abysmal working conditions in power looms, sickness of mills and the problems of mill workers have been identified as the issues that have not been addressed sufficiently

Table 1.1 Major policy events in Indian textile sector since independence

Policy event	Year
Reservation of some products under handlooms	1950
Cess on mill cloth	1952
Kanungo committee favouring power looms over handlooms for efficiency	1952
Karve committee: freezing mill output except exports; Asoka Mehta committee:	1955
promoting power looms establishment of national textile corporation	1964
Setting up of Cotton Corporation of India (CCI) to control cotton prices	1968
B. Sivaraman's study team: benefits meant for handlooms are reaped by power looms	1971
removal of excise duties for hank yarns for handlooms	1974
Soft loan scheme	1975
Freezing of power loom output to existing levels	1975
Removal of price control of cloth	1978
Abandoning of CCI and its buffer stock operations (except in Maharashtra)	1978
Textile policy: many reforms favouring mill sector, power looms and non-cotton	1981
textiles	1985
Technical Upgradation Fund Scheme (TUFS) and deregulation	1991
Dereservation of garment sector from small-scale industries sector phasing	2000
out of MFA quotas	2005

by the 1985 textile policy. Area-based approach and the establishment of new institutional arrangements such as an apex council have been recommended.

The textile policy framework of India has been motivated by five major issues: regulation of inter-sectoral competition, provision of cheap cloth, fibre policy, modernization of the industry and sickness and rehabilitation of mills. Table 1.1 summarizes some major policy events in Indian textile sector, based on the description by Siebert et al. (1980). Thus, the policy has been aimed at learning from the bad experiences of the British Raj and ushering the industry into a new era. However, major developments in the industry happened only in the late twentieth century (Table 1.1).

The mill sector has lost its market share in the domestic market, while the share of the power loom sector has been rising steadily due to lower production costs. While the weak mills have been suffering from many ills, others have prospered and diversified production. The power loom sector, taking advantage of its cost structure, black market, low fixed capital cost per unit of output and avoidance of excise duty, has prospered, though handloom sector is still increasing its production. Such trends in the structure of the Indian textile industry have affected the exports of fabrics and garments as well, owing to forward linkages (Uchikawa 1998).

A considerable trend in the textile sector in India in the recent years indicates that it has finally transformed enough away from its past. Employment growth has flattened relative to output and capital growth (see Table 3.2). This was not the case in the past few centuries since industrialisation, owing to the labour-intensive characteristic of the industry. A positive symptom for the textile workers in the recent past, however, is that the real wages are rising. On the other hand, this is not so good for the industrialists, who get their margins reduced owing to the rising wages. Given that the industry has remained in small-scale historically, one way of tackling this is increasing the scale of production. This has happened in the recent years (Narayanan 2008b).

Chapter 2
Description of India's Textile and Apparel Sector

A fibre is a particle, whose length is much greater than its thickness. Yarn is a continuous, often plied, strand composed of either fibres or filaments. A fabric is a material which is pliable, in short. All the processes involved in manufacturing fabrics from a bunch of fibres can be collectively termed as "textile industry". The process of converting a bunch of fibres into yarns is called spinning. For cotton, this starts from obtaining 'bale', which is a compressed mass of cotton fibres, prepared by a ginning factory, where cotton fibres are separated from their seeds mechanically. Its weight is about 170 kilograms. Bales are fed to opening and cleaning machines in the 'blowrooms' to remove foreign particles, dirt and dust, collectively called 'trash'. The cleaned mass of fibres is then 'carded' to aligned strands of fibres. A saying about spinning goes thus: well-carded is half-spun.

Subsequently, 'combing' could be done to remove short fibres called 'noil', in case yarns of higher quality, typically for hosieries, are required to be produced. 'Drawing' of the carded or combed 'sliver', which is the mass of aligned fibres from carding or combing process, is done to introduce some draft in the fibres. The twisting process starts with the 'simplex' machine, whose input is sliver and output is a twisted mass of fibres called 'roving', wound around simplex 'bobbins'. The final twisting and winding into small cops take place in the ring frames. The yarn from ring frame could then wind in cones, using the cone-winding machines.

Combing, drawing and simplex could be avoided in relatively new machines called rotor machines, which convert carded sliver directly into yarn. However, this can be used only for coarser yarns, with fineness less than 10s English Count,[1] which is the number of 840 yards in 1 pound of yarn. Hence, 10s means that there are 8400 yards in a pound of yarn.

There are many advancements in the processes of opening, cleaning, carding, combing, and drawing simplex and ring frames. A few of them have to do with

---

[1] English count is the number of 840 yards in 1 pound of yarn. Hence, 10s means that there are 8400 yards in a pound of yarn.

© The Author(s) 2018
B.N. Gopalakrishnan, *Economic and Environmental Policy Issues in Indian Textile and Apparel Industries*, SpringerBriefs in Environmental Science, DOI 10.1007/978-3-319-62344-3_2

automation, thanks to advanced electronic devices, and a few of them are concerned with higher quality of output and efficiency of operation. A few recent developments include electronically controlled devices in blowrooms by companies such as Trutzschler and Lakshmi Machine Works, compact spinning by Rieter Machine Works and fully automatic Autoconers by companies such as Schlafhorst, Savio and Muratec. A miniscule part of yarn production in India is hand-spun. Thus, the spinning sector of textile industry itself is heterogeneous, consisting of *khadi*, mills operating with obsolete machinery, mills operating with modern machines and those with state-of-the-art machinery.

Weaving is one of the techniques of converting yarn into fabric. This could be done using handlooms, power looms, semi-automatic looms, fully automatic looms or shuttleless looms, which are characterized by the absence of shuttle that carries pirns containing weft yarn, thereby increasing the speed and efficiency of power loom by leaps and bounds. In all of these, the basic concept is one of inserting (called 'picking') a 'weft' yarn in between two lifted up (called 'shedding') and down alternately and then pushing the 'weft' yarn (called 'beating up'). Warp and weft are just technical terms given to the different yarns involved in this process. This involves many preparatory processes such as warping, sizing, pirn winding and even yarn dyeing at times. A detailed account of weaving process is given in Niranjana and Vinayan (2001) or any basic book on textile technology.

Handlooms are of two types: throw-shuttle and fly-shuttle pit looms. In a throw-shuttle pit loom, the weaver has to sit on the edge of a small pit, operate the treadles to move the reed up and down and throw the shuttle across the width of the loom by hand. Fly-shuttle pit loom, which was introduced in 1920s and 1930s, contained strings tied with the shuttle driving cocks on both sides. In addition to cotton, rayon, polyester (both pure and blend) and silk are also used in handlooms for the past few decades. There are many independent handloom weavers in India, a majority of whom are affiliated to government-funded co-operative societies that operate in most states.

The recent developments in weaving include the evolution of shuttleless looms. Air-jet looms, water-jet looms, projectile looms and rapier looms are the examples of shuttleless looms. In India, there are many composite mills, which undertake both spinning and weaving operations. Knitting is another process of converting yarn into fabrics. This is done to result in a relatively more elastic and flexible fabrics, which are mostly used for manufacturing casual wear and innerwear. This is also a growing sector, in terms of modernisation. After the manufacture of fabrics, they need to be processed to improve their aesthetics as well as durability. There are numerous processes involved in textile wet processing, such as scouring, bleaching, dyeing, printing, mercerisation, etc.

All the processes involved in manufacturing garments other than those included in the textile sector, which is elucidated above, come under apparel or garment sector. This involves cutting of fabrics, based on the patterns used for stitching, and sewing the cut fabrics to manufacture apparel/garments, finishing of garments, pressing and packing. This sector was reserved for small sector in India till 2000, and hence this is predominantly carried out in very small scales. There are a few big companies such as

Gokaldas Creations, Madura Garments, Eastman Exports and Centwin Exports in India but even their scales are too low, compared to those in countries such as China. Exports from this sector gained momentum in the 1970s and this is a key export item for India. Advancements in this sector involve automation, enhanced efficiency, computerization and high quality ensured by better monitoring and control.

Silk industry depends on sericulture, which is an allied agricultural activity, as it involves rearing silkworms by feeding, typically, mulberry leaves, which are obtained from the farm. The filament that is secreted by the silkworms is then reeled, twisted and processed, for further production. There are new advancements such as texturization of silk, which enhance its moisture absorption capacity as well as soft feel, though it reduces its lustre, to some extent. In India, silk production is mostly carried out as a cottage industry, though there are a few big players such as Bharat Silks, which are big and export-oriented.

The way silk filaments are secreted from the mouth of silkworm is probably the inspiration for the design of spinnerets, which are huge plates with tiny pores in them, over which the chemicals in liquid form are poured to obtain synthetic filaments. These are then directly twisted into the so-called filament yarns or cut into fibres and then spun into the so-called staple fibre yarns. This is how most synthetic fibre-based yarns, such as nylon, polyester, viscose and acrylic, are manufactured. While viscose chemically as well as physically resembles silk, nylon and acrylic resembles wool and polyester resembles cotton.

The affinities between certain sets of fibres, coupled with the possibility of obtaining desired qualities of each of them, have encouraged the processes called mixing and blending. These are only subtly different, in the sense that mixing is done physically in fibrous state, while blending is done chemically in liquid or powder state. For example, polyester staple fibres are mixed with cotton staple fibres, in order to get a cheap, durable, resilient and comfortable mix of polyester-cotton fibres. Man-made fibre industry as well as manufacture of mixed fibres are expanding rapidly in India, despite strong excise controls, which are coming down over years. Reliance Group, Indorama and Sanghi Spinners are well-known polyester manufacturing companies in India.

There are various other subsectors of the textile sector. Manufacture of nonwovens and composites is an emerging area in most developed countries, while it is till at its infancy in India. Industrial textiles are the specially manufactured textile fabrics and yarns that are used in industries. Examples are nylon cords used in rubber tyres, strong and coarse fabrics used for reinforcement in many mechanical devices, fibres, etc. Geotextiles are a type of textiles that are used to reinforce the roads by a fabric layer beneath the surface of the earth, as done in Konkan railway route to ensure protection from landslides and soil erosion. Intelligent textiles are the new-generation electronics-integrated and/or high-technology textile materials that can perform functions such as camouflaging, changing the temperature and phase depending on various external parameters, power generation, entertainment, etc. Phase-changing materials are widely used for these purposes, based on advanced techniques based on physics and chemistry. There are numerous other applications of textiles, with limited scope, such as fancy yarns, ribbon fabrics, etc.

Chapter 3
Structure of India's Textile and Apparel Sector

3.1 Introduction

The Indian textile[1] and apparel sector[2] is the second largest employer after agriculture, with more than 33 million persons engaged in it. It contributes 1% to the gross domestic product (GDP), 15% to the total exports and 8% to the total manufacturing output of India, in 2004–2005 (based on calculations from Annual Survey of Industries and Directorate General of Foreign Trade, India). By virtue of being among the earliest established industries in the country and being a major sector responsible for rapid growth of the newly industrialized countries, in addition to the facts and figures listed above, textile industry is very significant for the Indian economy. This industry has a rich history in India, in addition to its dimensions in culture and heritage, so much so that any study on Indian history would be incomplete without a detailed treatment of textile history in India. Textile production has been an integral part of the lives of millions of poor people, including farmers in India, for centuries.[3] In addition, textile production has backward linkages with agriculture and allied activities, at least in the case of natural fibres.

Strong and diverse raw material base, cheap labour, ever-growing domestic market and relatively better technologies[4] than some of the other developing countries are the key strengths of the Indian textile sector that have resulted in such a

---

[1] The textile sector includes spinning that involves producing yarn from fibres, weaving that involves manufacturing fabric from the yarns and processing that involves chemical treatment and colouration of yarns and fabrics for durability as well as aesthetics.

[2] The apparel sector includes the processes that result in the manufacture of readymade garments from fabrics.

[3] Roy (1996) is a comprehensive study of Indian textile history.

[4] For example, Lakshmi Machine Works, India, is one of the largest textile machinery manufacturers in the world. Presence of companies like these has ensured that many advanced technologies are accessible to Indian industry.

© The Author(s) 2018
B.N. Gopalakrishnan, *Economic and Environmental Policy Issues in Indian Textile and Apparel Industries*, SpringerBriefs in Environmental Science,
DOI 10.1007/978-3-319-62344-3_3

pronounced prominence of this industry. Development of modern textile industry in India had gained momentum after it did so in Britain owing to availability of indigenous cotton, cheap labour, access to British machinery and a well-developed mercantile tradition in India.

The coexistence of a broad spectrum of production techniques, a distinct trend towards decentralized manufacture in the informal sector, sustained, albeit considerably declined, predominance of cotton as the raw material, a very huge sick public sector, a recent trend of the manufacturers of adopting the modern techniques, and existence of quite a few regulations and preferential tariff structure (favouring natural fibres and conventional means of production) are some fundamental features of the Indian textile and clothing industry in brief.[5]

Despite being one among the leaders in textile production in 1950 and the fact that India has a self-reliant value chain of textiles, India had been steadily receding from the world textile market, with a loss of importance in industrialization at home also. The decline of the Indian textile industry is very conspicuous relative to the other industries as well as relative to the textile industries of the other countries in the developing world as evident from a steep fall in the share of Indian textiles in the international market and that in the total Indian exports.

In the 1990s, the Indian textile industry had been facing a severe recession in terms of employment as well as the number of operational mills/factories, which continued despite fundamental changes in tariff structure among other policy aspects in the mid-1980s and in 1991. Though there were symptoms of late recovery, owing to the market expansion resulting from the phasing out of MFA quotas, there has been an astonishing decline in exports growth in 2006–2007 to 10.53% from over 16% in 2005–2006 (Ministry of Textiles 2007).

However, the objective of this chapter is not to examine the performance of Indian textile and apparel sector in international trade. Rather it focuses on some of the major domestic issues that encompass supply and demand in this sector. As regards the supply side, performance and employment in organized and unorganized segments are considered separately. The key aspects that are analysed are partial productivity measures, employment, capital and output. As for the demand side, the focus is on the fiscal and tariff policies in textile and apparel sector and their implications for demand.

This chapter is divided into six sections, in addition to this introduction. Section 3.2 gives an account of performance of the organized textile sector in India. Section 3.3 analyses the performance of unorganized textile sector. Consumption of textiles by Indians and the factors affecting it are analysed in Sect. 3.4. Section 3.5 concludes and elucidates some policy inferences of the analysis in this chapter.

---

[5] Misra (1993) and Sastry (1984) elaborate on these issues.

3.2   India's Organized Textile Sector: Performance and Employment

During the past decades, numerous textile mills have been closed and declared sick, while many of the mills under National Textile Corporation (NTC) continued to operate, despite losses, owing to the fact that there are many employees involved. Even in the private sector mills, employment had been a major issue. Though the sector has recovered to a large extent, its performance post-MFA has not been encouraging.

A wide range of regulations in the textile industry involving bureaucratic difficulties in the expansion of the industry and highly distortionary tariff structure were partly responsible for this steady recession in the past. For example, hank yarn obligation[6] required the spinners to allocate a fixed part of their production to handloom weavers. This not only restricted the profits of spinners but also the raw material access and cost for the weavers and others up the value chain. The reservation of garment sector[7] under the small-scale industry had restricted large-scale investment in this sector, which led to huge loss in efficiency that could have been otherwise achieved by economies of scale.

In the informal or unorganized sector that is progressing well in the clothing sector, the processes are not planned and systematic. The working conditions are not satisfactory as the labour regulations cannot be enforced and a hire-and-fire principle is in place. This is true even in a part of organized sector, wherein the manufacturers recruit contract labourers in order to minimize the losses that they are facing due to the inflexible labour regulations that stop them from firing their permanent employees even during recessions. In fact, some studies observe a rapid growth of the informal sector in the textile industry, especially after the reforms of 1991.

Table 3.1 shows the trends in annual average growth rates of some major variables for the aggregate textile industry. Since this was based on the aggregated textile data, figures could be calculated for four decades with proper concordance of different

Table 3.1 Average annual growth rates in the organized textile and apparel sector in India (1993–1994 prices)

Period	Output	Employment	Real wages	Real fixed capital
1961–1962 to 1970–1971	5.034	0.496	2.487	3.645
1971–1972 to 1980–1981	6.668	3.295	2.882	4.643
1981–1982 to 1990–1991	8.174	−0.968	5.44	8.802
1991–1992 to 1999–2000	6.718	0.997	2.378	17.774
1980–1981 to 1997–1998	5.34	−5.17	5.35	8.11
2001–2002 to 2004–2005	8.90	4.79	5.18	2.73

Source: Author's calculations from annual survey of industries

---

[6] It came into place in 1974; it was fixed at 50% of the total marketable yarn in 1986, though it was brought down to 40% and then 20% in 2003.

[7] This has been withdrawn with effect from November 2, 2000.

reports of Annual Survey of Industries. It can be seen that output, wages and fixed capital have been growing at an increasing rate from 1961–1962 to 1999–2000 but for a small fall in growth rate from 1991–1992 to 1999–2000.[8] The trend in the growth of employment is, however, not uniform. But for the period between 1971–1972 and 1980–1981, it has grown at a much lower rate than the other variables in most periods and in fact has declined from 1981–1982 to 1990–1991.

Though employment grew on an average after the reforms of 1991, its growth was nowhere comparable to the growth of the other variables, especially, capital stock, which has grown at about 18% a year.[9] This observation is even more precise if the period from 1980–1981 to 1997–1998 alone is examined, since, in this period, employment has fallen at approximately an annual average rate at which output has grown, despite a remarkable annual growth of capital of over 8%. It would seem from this that, as a whole, textile industry is characterized by substitutability between capital and labour. Given the labour-intensive nature and unionized labour of the organized segment of this industry, entrepreneurs might have had capital to substitute the labour. Even then, the absolute fall of 5 per cent per year in the employment when output has increased by 5 per cent per year draws attention.

Another striking observation from this table is that the recent years (2000–2001 to 2004–2005) have seen sharpest growth in organized sector employment. This is seen along with a decent growth in real wages and output. This rise in employment is despite the growth of unorganized sector and contract workers within organized sector, to face stiff cost competition mainly in the wake of gradual phasing out of quotas in this period. The growth in capital has come down to below 3%, which is another reason to worry since to face the competitive market in the free-trade regime, huge investment is required to be poured in.

Three measures of partial productivity have been analysed in Table 3.2: capital productivity, capital intensity and labour productivity. Capital productivity is the ratio of gross output to gross capital. This gives the amount of output produced from a unit of capital. Capital intensity is defined as the ratio of gross capital to total employment. This reflects the relative size of capital and labour in the industries. Labour productivity is the ratio of gross output to total employment. This measures the extent to which labour has been used for production.

Table 3.2, in terms of Lakhs (1 Lakh = 0.1 Million) of rupees of gross value of output and gross invested capital per person engaged, makes it more explicit that the textile industry, on an average, has precisely become much less labour-intensive than it was 30 years ago. A rise in capital-labour ratio despite a fall in capital productivity seems to suggest an existence of mere substitution of labour by capital, at

---

[8]This might partly be due to the omission of cotton ginning sector for the two years after 1997–1998, as the NIC-98 (National Industrial Classification-1998) has classified this sector under agriculture, while the pre-1997–1998 data is based on NIC-87 (National Industrial Classification-1987). The same argument holds for the other variables also, and hence the figures for the period between 1980–1981 and 1997–1998 have been highlighted.

[9]This is quite as expected, since this was the period when the phasing out of MFA quotas was initiated, and hence the firms were apparently getting ready for the free-trade regime by attempting to invest and enhance their quality and scales as well as the consequent economies and efficiency.

Table 3.2 Trends in some ratios of capital (K), output (Y) and employment (N)

Year	Y/K	K/N	Y/N
1973–1974	2.569	4.523	11.616
1980–1981	3.657	4.364	15.958
1985–1986	3.092	7.331	22.664
1990–1991	3.614	10.332	37.336
1997–1998	1.546	34.122	52.76
2001–2002	1.403	3.969	6.443
2004–2005	1.777	4.426	7.864

Source: Author's calculations from annual survey of industries

least till 1997–1998. By 2001–2002, capital productivity, capital intensity and labour productivity have fallen sharply. There has been a gradual in all these measures by 2004–2005. This is a serious problem, given the fact that the international market is becoming more and more competitive, requiring high productivity and capital intensity.

Capital productivity (Y/K) has been quite stable from the seventies till 2005, varying between 1.4 and 3.7. However, there are bulges in capital intensity (K/N) as well as labour productivity (Y/N). Strikingly huge values for these values during 1985–1986, 1990–1991 and 1997–1998 could possibly be a result of rapid fall in employment, which is in the denominator for both these measures, in this period, as can be inferred from Table 3.1. Growth of employment by 2001–2002 might have offset the unusually high rise in these ratios before, hence explaining the fall in these ratios to much lower values. However, a not-so-high growth of capital since 2001–2002 has led to increase in capital productivity by 2004–2005, while impressive output growth rate has caused a rise in both capital and labour productivity.

In the recent years, most of the measures of protection have been brought down as a part of the reforms. Table 3.3 shows effective rates of protection for different subsectors of textile industry over few years of the past. The measure used is based on Das (2003), who defines effective rate of protection as a measure of the extent to which sector is sheltered from foreign competition. Specifically, this is based on Corden's formula and is the percentage excess of domestic value added, vis-à-vis world value added, introduced because of tariff and other trade barriers. This measures the distortion introduced due to tariff on the input prices as well as the final output prices and therefore measures protection to domestic factors of production. We use this measure of protection, because this not only captures absolute level of effective rate of protection of each sector but also accounts for the inter-sectoral differences in protection mentioned above. It is evident from this table that protection has fallen in all subsectors, and this reduction has been strikingly sharp in cotton khadi and handlooms. Fall in protection may have implications for employment to the extent that protected industries that tend to lose because of fall in protection are employment-intensive.

Table 3.3 Trends in effective rates of protection for different subsectors in Indian textile sector

NIC-1987 Codes	Description of sectors	1980–1985	1986–1990	1991–1995	1996–2000
230,231,235	Cotton ginning, spinning and weaving	109.77	125.38	68.38	42.93
262	Embroidery, ornamental trimming and Zari	160.91	151.23	95.79	48.22
232,233	Cotton khadi and handlooms	109.36	126.85	70.95	0
234,236	Power looms and processing in mills	109.77	125.38	68.38	42.93
260,265,267	Hosieries, garments and other made-ups	138.33	149.89	98.45	54.25
263	Carpets and other furnishings	102.52	91.8	63.3	44.66
268,269	Water-proof and other specialty textiles	160.91	151.23	95.79	48.2

Source: Das (2003, ICRIER Working Paper No:105)

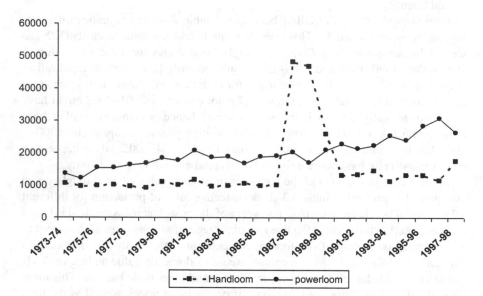

Fig. 3.1 Employment trends in non-mill textile sector (*Source*: Author's calculations from Annual Survey of Industries; *Note*: Post-1997–1998 data exists in NIC-98 classification, which does not allow us to look at the sectors in this scheme of disaggregation, done using NIC-87. So, this analysis stops with 1997–1998)

It is useful to examine the employment trends in some subsectors, using the past data, and link them with some policy measures. Figure 3.1 shows that employment in handlooms and power looms is more or less stagnant from 1973–1974 to 1997–1998, except for the fact that the Handlooms (Reservation of Articles for Production) Act of 1985, which was enforced from 1986, had caused a sharp increase employment in

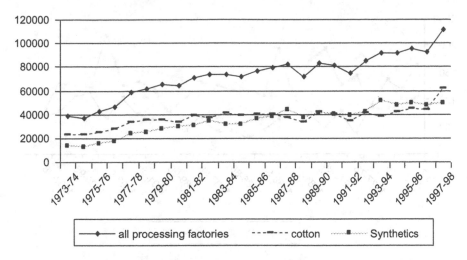

Fig. 3.2 Employment trends in different subsectors of textile wet processing sector (*Source*: Author's calculations from Annual Survey of Industries; *Note*: Post-1997–1998 data exists in NIC-98 classification, which does not allow us to look at the sectors in this scheme of disaggregation, done using NIC-87. So, this analysis stops with 1997–1998)

handloom sector in 1986–1987. However, this fell rapidly owing to the liberalization that favoured the power looms and mill sector in the late 1980s, leading to the past levels of employment. Figure 3.2 shows that employment has been consistently falling in the cotton mill sector, while it has been almost stagnant in the wool, silk and other natural fibres and risen sharply in the synthetics and made-up textiles, more so after the reforms of 1991. This roughly indicates that the highly regulated conventional cotton mill sector has suffered the most among all the subsectors of the cotton textiles in terms of employment, implying the existence of a negative relationship between labour regulations and employment. This also suggests a positive effect of liberalization at least in some subsectors that come under the made-ups.

Figure 3.3 shows that though employment has been rising as a whole in the textile processing sectors that are prime polluters in the industry, its fall in 1987–1988 and 1995–1996 in the overall, cotton and synthetic processing sectors indicates the possible existence of a negative impact, at least in short term, of the Environmental Pollution Act (1987) and the ban of certain dyes by some members of the European Union in 1995. Figure 3.4 strengthens evidence for this statement since fall in employment is even more conspicuous in the case of wool and silk processing sectors, which are more pollution-intensive in nature. Despite all these short-term trends, the long-term increasing trend is still preserved, suggesting that the rise in employment that might be gained by compliance to these regulations may have played a role in increasing employment. This fact is also confirmed by a rigorous econometric exercise based on a comprehensive theoretical framework by Narayanan (2007a) and Narayanan (2005b).

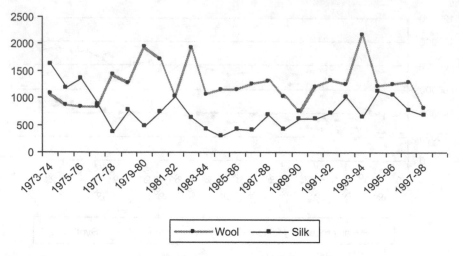

Fig. 3.3 Employment trends in textile wet processing sector (*Source*: Author's calculations from Annual Survey of Industries; *Note*: Post-1997–1998 data exists in NIC-98 classification, which does not allow us to look at the sectors in this scheme of disaggregation, done using NIC-87. So, this analysis stops with 1997–1998)

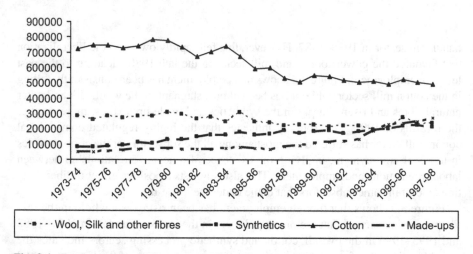

Fig. 3.4 Employment trends in subsectors in textile sector (*Source*: Author's calculations from Annual Survey of Industries; *Note*: Post-1997–1998 data exists in NIC-98 classification, which does not allow us to look at the sectors in this scheme of disaggregation, done using NIC-87. So, this analysis stops with 1997–1998)

Table 3.4 shows that the number of factories has increased in both textile and apparel sectors, implying a spectacular rise in fixed capital, number of workers, total persons engaged, total emoluments and gross output in apparel sector. The growth in most of the variables has not been so high in textile sector. However, growth in number of factories in apparel sector has been less than half that in textile sector, highlighting

Table 3.4 Salient features of the organized textile and apparel sector in India: recent trends (values are in Rs.; Lakhs current prices and others are in number)

Year	Sector	Factories	Fixed capital	Workers	Total persons engaged	Wages to workers	Total emoluments	Gross output
2001–2002	Textile	12,557	3,931,489	1,004,848	1,182,124	445,017	602,216	8,202,046
2001–2002	Apparel	3283	310,821	272,524	317,089	86,647	127,917	1,456,746
2001–2002	Total	15,840	4,242,310	1,277,372	1,499,213	531,664	730,133	9,658,792
2004–2005	Textile	13,521	4,529,094	1,076,480	1,264,427	480,456	714,951	11,108,327
2004–2005	Apparel	3397	470,132	387,606	450,175	139,024	222,732	2,374,789
2004–2005	Total	16,918	5,099,226	1,464,086	1,714,602	619,480	937,683	13,483,116
Average annual growth rates								
2001–2005	Textile	2.559	5.915	2.376	2.321	2.655	6.240	11.811
2001–2005	Apparel	1.157	17.085	14.076	13.990	20.150	24.707	21.007
2001–2005	Total	2.269	6.733	4.872	4.789	5.506	9.475	13.198

Source: Author's calculations from Annual Survey of Industries (2001–2002 and 2002–2003)

the fact that there is a massive consolidation and scaling up taking place among apparel manufacturing enterprises. Partly, this could be attributed to the fact that garment sector was dereserved from the small-scale industries sector in 2000 and the resultant mergers of smaller fragments after dereservation, causing an actual reduction in number of factories, which could have been outweighed by the number of new factories established.[10] Thus, organized apparel sector appears to be getting more tuned for the free-trade regime than the organized textile sector, in terms of scaling up, though the latter has always been a large-scale sector compared to the former.

Higher growth in number of workers than in total employment, coupled with the conspicuously higher growth rates in total emoluments than in wages, indicates that despite an increased demand for production workers, the pay increases are becoming higher for supervisory and managerial staff, in other words, more skilled employees.

However, a word of caution is to be in place while mentioning about employment in organized textile sector. Given the high labour costs and rigidities in the labour markets, coupled with the sickness of factories, the employers go for subcontracting employees from the unorganized sector, thereby reducing the employment in the organized sector. This, in addition to showing up as a decline in employment, is not a healthy trend, as far as the welfare of employees are concerned, as they are not protected by any legislation, given their unorganized nature. This needs to be taken care of by the policymakers, possibly by ensuring income security for the workers, coupled with some labour flexibility for the employers, so that they are discouraged to go for subcontracting.

Having analysed the trends in employment in India's organized textile sector, it is essential to link these observations with a perspective of development. The apparel sector has performed quite well in terms of employment in the recent years, showing a recovery from the decline in the past, though the same is not true in the case of textile sector, of course, with some signs of recovery. This seems to be a good indication for the country's development in general, given the immense contribution of textile sector to the economy. The story on employment and performance of textile industry would not be complete without a comprehensive examination of trends in unorganized textile sector. Hence, Sect. 3.4 analyses the performance, in terms of partial productivity trends as well as a few other factors.

3.3 Performance of India's Unorganized Textile Sector

In India, unorganized manufacturing sector is defined as the collection of those manufacturing units whose activity does not come under any statutory act or legal provision and/or which do not maintain any regular accounts or which are not

---

[10] It may be noted here that an investment up to Rs 3 crore in plant and machinery and an FDI-cap of 25 per cent is permitted, subject to an export obligation of 50 percent of total production of garments, even before dereservation.

Table 3.5   Shares of various subsectors in different sectors for the year 2000–2001 (current prices)

Subsector	Sector	% Share in gross value added	% Share in employment
Unorganized manufacturing	Total manufacturing	28	73
Unorganized textiles	Total textiles	18	74
Unorganized apparel	Total apparel	59	89
Unorganized textiles and apparel	Total textiles and apparel	31	79
Unorganized textiles and apparel	Unorganized manufacturing	29	31

Source: Author's calculations from Annual Survey of Industries and NSSO (2000–2001)

registered under Sects. 2 m(i)[11] and 2 m(ii)[12] of the Factories Act, 1948 and which are registered under Sect. 85[13] of Factories Act, 1948. As Table 3.5 reveals, unorganized manufacturing sector contributes 28% of the gross value added and 73% of employment to the total manufacturing including the organized sector, thus playing a vital role in Indian economy.

As Table 3.5 illustrates, the unorganized textile and apparel sector comprises 31% of gross value added and 79% of employment in the entire textile and apparel sector in India. In fact, the unorganized apparel sector, which contributes about 59% to gross value added and 89% to employment in the apparel sector in India, is predominantly unorganized. Thus, any study of Indian textile industry cannot claim completeness unless it considers unorganized sector in its analysis.

Misra (1993) notes that unorganized segment of India's textile sector comprises handlooms, power looms, small power-processors and traditional hand-processors, in addition to the numerous small-scale garment firms in woven as well as hosiery sectors. Power looms either operate on an independent basis or serve a master-weaver system, in which they just process the orders from the master-weaver providing the raw materials and charges based on quantity of cloth produced. They acquire loans from non-bank sources, while handlooms in rural areas rely on non-institutional sources such as village money lenders, unlike the organized weaving mills, at a higher rate of interest and from undeclared, untaxed and often illegal income.

In the urban areas, where this sector is dominant, the labour is mostly the migration from the rural areas, non-unionized, and hence is obtained at market-determined wage rates much lower than organized sector. All these, in addition to exemption of grey fabric from excise duty and sales taxes and long working hours, are the sources of competitive advantage for unorganized power loom sector, over the organized mill sector. In fact, the rapid growth of power loom sector after deregulatory

---

[11] Factories using power and employing ten or more workers on any working day.

[12] Factories not using power and employing 20 or more workers on any working day.

[13] Factories, which have less than 10/20 workers with or without power, specially notified by state government.

measures introduced in 1985 could be attributed to its unorganized labour market, well-developed input markets, ease of entry and flexible specialization.

Although there are some large production centres of handlooms in urban areas, major part of this sector is small-scale, often as an ancillary activity to agriculture in rural areas. Many of the Indian handlooms are even non-commercial, such as those in the North East, which produce for local or domestic consumption. There are small-scale power-processors as well as hand-processors using traditional techniques in India. The fact that the raw material cotton cost is around one-fourth of the total value, and the three stages of spinning, grey weaving and processing each progressively add one-fourth of final value, illustrates well the importance of processing and weaving in the cotton textile value chain.

Further down the value chain, most of knitted garment manufacturers are in the unorganized sector. For example, many firms in Tiruppur, an industrial town in Tamil Nadu, are either unorganized or depend heavily on subcontracting to firms in unorganized sector. Most of these firms are export-oriented and are seasonal/casual in operation, depending on the orders from the foreign buyers. These firms are usually specialized[14] and small, and hence they get their job orders[15] done with the help of numerous suppliers. Even some of the woven garment manufacturers, such as a few in Mumbai, Gurgaon, Chennai and Bangalore, are unorganized.

It is worth mentioning that the aforementioned characteristics are almost typical for the cotton sector. However, the features of the other sectors such as wool, silk and synthetics, which involve similar processes, remain the same. Jute sector, which is concentrated in rural and urban areas of West Bengal, among few other states, has gone through major transformation from prosperity pre-independence to illness in recent years. Coir sector is a major cottage industry in many rural areas in Tamil Nadu and Kerala. Other miscellaneous sectors include furnishings; manufacture textiles for industrial purposes such as Nylon tyre cords, metallized yarns and rubber thread or cord covered with textile material; speciality textiles such as tapes, cords and nets; and fancy textiles such as embroidery, zary work, and wadded textiles.

As the Multi-Fiber Arrangement (MFA) quotas are being phased out, Indian textile sector is facing both opportunities and threats. While the organized segment of the sector seems poised for a boom owing to its relatively better economies of scale, the large unorganized sector in this industry is expected to suffer because of its lack of competitiveness and technical efficiency among other related factors such as insufficient scales of operation that limits the level of efficiency and competitiveness that can be achieved by these firms.

Further, dereservation of the garment sector under the small-scale industries sector in 2000 is expected to have adverse effects on the unorganized sector, as the enterprises in this sector would now have to face stiff competition from big players that would enter the market with this development in the policy, which were already threat in the export sector, as the upper limit of investment was higher then, as footnoted elsewhere in this paper. Given the huge contribution of unorganized sector to

---

[14] Of course, there are a handful of firms that carry out all the activities in the textile value chain.

[15] Most firms are order-based, though there are few which market their products too.

Table 3.6 Annual average growth rates in unorganized textile sector (based on 1993–1994 prices)

Period	Employment	Fixed assets	Wages	Output
1984–1990	−11.803	−24.19	−8.787	−24.512
1989–1995	2.724	−8.412	9.174	−3.276
1994–2001	6.781	−9.123	10.946	−7.251

Source: Author's calculations based on NSSO (1989, 1994, 1998, 2002)

the textile sector, this is certainly a serious issue for this sector as a whole. On the other hand, small firms are competitive after the recent trade reforms, as decentralized production does have some strengths in terms of costs. In addition, mergers of smaller firms into bigger ones could be a solution to face the competition from big players. Combined effluent treatment plants established in clusters of small textile dyeing units, in places like Tiruppur, are examples of how the small firms can join hands to eliminate their disadvantage of lack of economies of scale.

Given the heterogeneity of the unorganized textile sector, coupled with their potential strengths and drawbacks, both as explained in the previous paragraphs, it is imperative to examine the trends in productivity in this sector in the recent years for which detailed data is available.

This analysis uses the aggregate summary results of 40th, 45th, 51st and 56th rounds on unorganized manufacturing of the National Sample Survey Organization (NSSO) (NSSO 1989; NSSO 1994; NSSO 1998; NSSO 2002). The different types of enterprises in this study are own account manufacturing enterprises (OAME) consisting of no employee other than the working owner, non-directory manufacturing establishments (NDME) employing less than six persons other than working owner and directory manufacturing establishments (DME) employing more than six persons other than working owner.

Based on this data,[16] we analyse the average annual growth rates in employment, fixed assets, wages and output, as illustrated in Table 3.6. While employment and wages have fallen, on an average, from 1984 to 1990, they have risen in the early nineties, and their growth has been much higher in the late 1990s. This is despite a fall in fixed assets and output throughout this period, though the decline has not been as high in 1990s as it was in 1980s. The interesting observation herein is that this trend is exactly the reverse of what has happened in organized sector: decline in employment despite rise in capital and output.

Partial productivity measures should be analysed to obtain an overview of the performance of unorganized textile sector. Here, we analyse three of them: capital productivity (no units), labour productivity and capital intensity (in rupees per employee). In large-scale or capital-intensive industries, capital productivity may be expected to be much lower than unity, as output produced would require a capital

---

[16] The demerits of NSSO data on unorganized manufacturing sector are possibility of unrepresentative sampling, response errors, inadequate sample size and absence of sampling error estimates. Owing to the absence of any better source of data for unorganized textile sector, we use this data for this analysis, acknowledging these limitations.

that is much higher than itself, owing to the capital-intensive nature of production. However, as we are considering unorganized sector, which is not very much likely to include such enterprises, this ratio may be even greater than one. This indicates the extent to which capital has been used for production.

As we construct these measures over years, to facilitate inter-temporal comparability, we obtained them in constant prices (base year, 1981–1982) by deflating the fixed assets using WPI for textile machineries and gross output using WPI for the respective products, namely, textiles and apparel.

Tables 3.7 and 3.8 show the trends and growth rates, respectively, in capital intensity, capital productivity and labour productivity across different enterprises and areas in the two subsectors of the textile sector, namely, textile manufacture (NIC-98 code, 17) and apparel manufacture (NIC-98 code, 18).

Firstly, we compare the trends in these variables for each year across different enterprise types, areas and subsectors. Secondly, we look at the average annual growth rates in them, during a few years in the past. Thirdly, we shall derive overall inferences from this analysis.

3.3.1 Capital Productivity

In 1984–1985, non-directory manufacturing enterprises (NDME) were more capital-productive than own-account manufacturing enterprises (OAME) in almost all categories, except in the rural textile sector, where both are comparable. While the urban textile NDME sector produces output that is more than thrice that of capital, output is as high as capital in most other sectors except apparel OAME. In all cases except rural NDME,[17] apparel sector is less capital-productive than textile sector. Rural textile NDME is the only exception for the observation that all categories in rural areas have been more capital-productive than those in urban areas.

In 1989–1990, all categories except rural textile OAME had capital productivity measuring above unity, exhibiting higher levels compared to those in 1984–1985, except urban textile NDME in which it has halved. Further, NDME are more capital-productive than OAME in all categories, thereby comprising four best ones among them in terms of capital productivity. All categories in rural areas have been more capital-productive than those in urban areas, except textile NDME, just as was the case in 1984–1985. Further, in all categories except urban NDME, apparel sector has been more capital-productive than textile sector.

In 1994–1995, DME is also including in the analysis, owing to the availability of its data from the same source (NSSO 1998). In this year, all categories in NDME, except urban apparel sector, were more capital-productive than OAME, while those in DME, except rural apparel sector, were better than those in NDME. Compared to 1989–1990, capital productivity has fallen in all categories except rural apparel NDME. While urban textile NDME had been most capital-productive of all categories

---

[17] Note that in this case, both textile and apparel sectors are equally capital-productive.

Table 3.7 Trends in partial productivity measures in unorganized textile sector in India

Year	Subsector	Sample	Enterprise type	Capital productivity	Capital intensity	Labour productivity
1984–1985	Textile	Rural	OAME	0.902	2016.479	1819.41
	Apparel	Rural	OAME	0.251	8600.825	2154.82
	Textile	Urban	OAME	0.687	3679.076	2527.268
	Apparel	Urban	OAME	0.108	39,475	4281.939
	Textile	Rural	NDME	0.863	5204.038	4488.943
	Apparel	Rural	NDME	0.884	4554.78	4026.569
	Textile	Urban	NDME	3.263	3648.323	11903.93
	Apparel	Urban	NDME	0.695	9940.026	6906.713
1989–1990	Textile	Rural	OAME	1.021	1742.425	1778.319
	Apparel	Rural	OAME	1.253	1879.168	2354.176
	Textile	Rural	NDME	1.74	2435.485	4238.551
	Apparel	Rural	NDME	1.757	3445.447	6054.648
	Textile	Urban	OAME	0.713	4247.893	3030.697
	Apparel	Urban	OAME	1.069	4832.785	5165.134
	Textile	Urban	NDME	1.871	10575.99	19787.06
	Apparel	Urban	NDME	1.303	12223.04	15922.37
1994–1995	Textile	Rural	OAME	1.143	2033.08	2323.994
	Apparel	Rural	OAME	1.166	1596.906	1862.392
	Textile	Urban	OAME	0.836	4524.921	3782.379
	Apparel	Urban	OAME	0.889	5100.408	4532.575
	Textile	Rural	NDME	1.279	4335.058	5542.978
	Apparel	Rural	NDME	2.31	1965.746	4541.748
	Textile	Urban	NDME	1.251	11294.63	14133.05
	Apparel	Urban	NDME	0.493	24059.05	11871.76
	Textile	Rural	DME	1.578	5905.005	9320.225
	Apparel	Rural	DME	2.244	3438.526	7717.756
	Textile	Urban	DME	1.804	9804.714	17688.04
	Apparel	Urban	DME	2.8	6893.022	19301.48
2000–2001	Textile	Rural	OAME	0.906	2577.797	2336.765
	Apparel	Rural	OAME	0.612	4986.596	3050.152
	Textile	Rural	NDME	1.16	4680.898	5429.882
	Apparel	Rural	NDME	0.794	6554.459	5202.371
	Textile	Rural	DME	1.575	6661.292	10490.51
	Apparel	Rural	DME	1.201	5341.884	6418.246
	Textile	Urban	OAME	0.653	6369.44	4159.148
	Apparel	Urban	OAME	0.43	10000.64	4296.527
	Textile	Urban	NDME	1.49	15329.54	22846.26
	Apparel	Urban	NDME	0.539	15875.74	8554.678
	Textile	Urban	DME	1.452	16719.66	24275.79
	Apparel	Urban	DME	1.049	16444.34	17243.16

Source: Author's calculations based on NSSO (1989, 1994, 1998, 2002)

Table 3.8 Growth trends of partial productivity measures in unorganized textile sector in India

Period	Subsegment	Sample	Enterprise type	Capital productivity	Capital intensity	Labour productivity
1984–1985 To 1989–1990	Textile	Rural	OAME	2.623	−2.718	−0.452
	Apparel	Rural	OAME	80.007	−15.63	1.85
	Textile	Rural	NDME	20.351	−10.64	−1.116
	Apparel	Rural	NDME	19.756	−4.871	10.073
	Textile	Urban	OAME	0.772	3.092	3.984
	Apparel	Urban	OAME	177.059	−17.551	4.125
	Textile	Urban	NDME	−8.532	37.977	13.245
	Apparel	Urban	NDME	17.495	4.594	26.107
1989–1990 To 1994–1995	Textile	Rural	OAME	2.4	3.336	6.137
	Apparel	Rural	OAME	−1.381	−3.004	−4.178
	Textile	Urban	OAME	−4.358	−1.274	−5.354
	Apparel	Urban	OAME	−3.37	1.108	−2.449
	Textile	Rural	NDME	−5.306	15.599	6.155
	Apparel	Rural	NDME	6.296	−8.589	−4.997
	Textile	Urban	NDME	−6.624	1.359	−5.715
	Apparel	Urban	NDME	−12.424	19.367	−5.088
1994–1995 To 2000–2001	Textile	Rural	OAME	−4.14	5.359	0.11
	Apparel	Rural	OAME	−9.51	42.453	12.755
	Textile	Urban	NDME	−1.856	1.596	−0.408
	Apparel	Urban	NDME	−13.129	46.687	2.909
	Textile	Rural	DME	−0.045	2.562	2.511
	Apparel	Rural	DME	−9.294	11.071	−3.368
	Textile	Rural	OAME	−4.376	8.153	1.992
	Apparel	Rural	OAME	−10.331	19.215	−1.042
	Textile	Rural	NDME	3.821	7.145	12.33
	Apparel	Rural	NDME	1.841	−6.803	−5.588
	Textile	Urban	DME	−3.904	14.105	7.449
	Apparel	Urban	DME	−12.511	27.713	−2.133

Source: Author's calculations based on NSSO (1989, 1994, 1998, 2002)

till 1989–1990, it has been just an average category in these terms in 1994–1995. Except in urban NDME, capital productivity has been higher in apparel sector than in textile sector, in all enterprise types and areas. Enterprises in urban areas have higher capital productivity than in rural areas only in DME and reverse holds true for other enterprise types.

In 2000–2001, capital productivity has conspicuously declined in all categories. All categories in DME, except urban textiles, are more capital-productive than others, while those in OAME are worse than those in others in this aspect. One striking observation is that capital productivity in apparel sector is lower than that in textile sector in all enterprise types and areas. In all cases except textile NDME, enterprises in rural areas are more capital-productive than those in urban areas.

As seen in Table 3.7, annual average growth rates of capital productivity from 1984–1985 to 1989–1990 were in two digits or even higher in all categories barring textile OAME, where they were less than 10%, and urban textile NDME, where it had fallen. From 1989–1990 to 1994–1995, average annual rates of decline in all categories, except textile OAME and apparel NDME in the rural sample,[18] range from 1% to 12%. Between 1994–1995 and 2000–2001, enterprises had been becoming 0.05%–13% less productive every year, on an average, except in the case of urban NDME, where they had become more productive at an average annual rate of 1.8% and 3.8%. These rates of decline were much higher in apparel sector than in textile sector. Even in urban NDME, apparel sector had become more productive at a rate lower than that at which textile sector had become. Decline in capital productivity, wherever it had occurred, had been more rapid in urban enterprises than in rural enterprises.

3.3.2   Capital Intensity

In 1984–1985, capital intensity varies between Rs. 2000 and Rs. 10,000 per employee, with an outlier of over Rs. 39,000 for urban apparel OAME sector. Capital intensity has been much higher in apparel sector than in textile sector, except in rural NDME, wherein it is the other way round. Except in textile NDME, the enterprises in urban areas are more capital-intensive than those in rural areas. With the exception of rural textile sector, NDME are less capital-intensive than OAME.

While these figures vary between Rs. 1700 and Rs. 12,000 in 1989–1990, enterprises in apparel sector, urban areas and NDME have been uniformly more capital-intensive than those in textiles sector, rural areas and OAME, respectively, with no exceptions. Except for the enterprises in urban textile OAME and urban NDME, capital intensity has fallen in all categories, the sharpest fall being more than eight times in the case of urban apparel OAME.

In 1994–1995, capital intensity ranged from Rs. 2000 to Rs. 24,000, and textile sector was more capital-intensive than apparel sector in the enterprises in rural areas and those in DME, though urban apparel NDME was most capital-intensive among all categories. Urban enterprises and NDME have been more capital-intensive than rural enterprises and OAME, respectively. While DME in rural areas were more capital-intensive than NDME in these areas, DME in urban areas have been less capital-intensive than NDME in these areas. Except rural apparel NDME, capital intensity has fallen in all categories in 1994–1995, compared to those in 1989–1990.

Unlike in 1994–1995, urban DME has been most capital-intensive (around Rs. 16,000, while the lowest is around Rs. 2600) category in 2000–2001, pushing urban NDME to second. Apparel sector has been more capital-intensive than textile sector in all categories except DME. OAME is less capital-intensive than NDME, which is less capital-intensive than DME, in all categories except rural apparel sector,

---

wherein DME is less capital-intensive than NDME. Further, we observe that enterprises in urban areas are much more capital-intensive than those in rural areas. Capital intensity is much higher during 2000–2001 than that during 1994–1995 in all categories.

Except urban NDME and textile urban OAME, enterprises in all categories had become less capital-intensive, at annual rates of 3%–18% from 1984–1985 to 1989–1990. However, annual growth rate has been as high as 38% in textile urban NDME. This decline in capital intensity could not be offset by growth in a few categories from 1989–1990 to 1994–1995, because rapid growth has been seen only in the categories which had, to begin with, grown in capital intensity from 1984–1985 and growth, if occurred, in the other categories were not high relative to the rates of decline in the previous period.

Unlike the previous periods, capital intensity had grown quite rapidly, in most categories, from 1994–1995 to 2000–2001, with the annual average growth rates ranging from 2% to 47%, the only exception being urban apparel DME. One more noteworthy observation is that apparel sector has grown capital-intensive much faster than textile sector, wherever it has grown, explaining why apparel sector has become more capital-intensive than textile sector in this year, in contrast with 1994–1995 figures. While growth rates had been much higher in textile sector in the urban sample than that in the rural sample, the reverse holds for the apparel sector, with an exception of DME. The other observations in growth rates may be made directly from Table 3.7.

3.3.3   Labour Productivity

While textile sector was less labour-productive than apparel sector in OAME, the reverse holds true for NDME, during 1984–1985. NDME were more labour-productive than OAME in all sectors and areas. Urban enterprises were more labour-productive than rural areas. While rural textile OAME was least labour-productive (Rs. 1800), urban textile NDME was most labour-productive (Around Rs.12000).

Except for rural textile enterprises, labour productivity had increased in all categories from 1984–1985 to 1989–1990. Urban enterprises and NDME were more labour-productive than rural enterprises and OAME, respectively, during 1989–1990. The fact that apparel sector was more labour-productive than textile sector is violated only by urban NDME, which is the most labour-productive (about Rs. 19,800). Rural textile OAME was least labour-productive, with about Rs. 1780 per person.

In 1994–1995, except in urban OAME and DME, labour productivity, which varied from around Rs. 1800 to Rs. 19,000, was less in apparel sector than in textile sector. Urban enterprises, DME and NDME were more labour-productive than rural enterprises, NDME and OAME, respectively.

During 2000–2001, DME were more labour-productive than NDME, which, in turn, were more labour-productive than OAME. With an exception of OAME, apparel sector was more labour-productive than textile sector. Urban enterprises

were more labour-productive than rural ones. Labour productivity varied from
Rs. 2300 to Rs. 24,000 during this year.

From 1984–1985 to 1989–1990, labour productivity grew in all categories at
average annual rates ranging from 1.8% to 26% except textile sector in rural sample,
where it has declined at relatively lower rates. In contrast, this had declined in all
categories except rural textile, in which it had grown at about 6% per year, from
1989–1990 to 1994–1995. This decline has been a bit more pronounced in apparel
sector than in textile sector.

In the period between 1994–1995 and 2000–2001, labour productivity has grown
in textile sector in all categories except rural NDME, in which it has declined at an
annual rate of less than 1%. In the rural areas, apparel sector had grown in this
aspect, at 3%–13% per year, except in DME, which saw a decline of around 3% per
year. Urban apparel enterprises have become less labour-productive in all categories
at 1%–6% per year.

Overall Inferences on Partial Productivity Measures
With a few exceptions, NDME, rural enterprises and textile sector were more
capital-productive than OAME, urban enterprises and apparel sector, respectively,
in 1984–1985. While capital productivity grew from 1984–1985 to 1989–1990 in
most categories, the other observations are the same as for 1984–1985, except that
apparel sector was more capital-productive than textile sector. From 1989–1990 to
1994–1995, capital productivity declined in almost all categories, with that of DME
being the highest among all enterprise types. The observation that DME in urban
areas are more capital-productive than those in rural areas is the only other difference between the figures in 1994–1995 vis-a-vis those in 1989–1990. In 2000–2001,
capital productivity has conspicuously declined in all categories, more so in urban
than in rural areas, explaining the fact that enterprises in rural areas were more
capital-productive than those in urban areas. One striking observation is the fall in
capital productivity in apparel sector both in absolute and relative terms, and hence
apparel sector was less capital-productive in apparel sector than in textile sector.

In 1984–1985, capital intensity was much higher in apparel sector, urban areas
and NDME than, respectively, in textile sector, rural areas and OAME with few
exceptions. The same is true for 1989–1990 with no exceptions, though capital
intensity fell sharply in most categories since 1984–1985. Between 1989–1990 and
1994–1995, there was little, no or negative growth in capital intensity.

Textile sector was more capital-intensive than apparel sector in rural DME. While
rural DME were more capital-intensive than rural NDME, urban DME were less
capital-intensive than urban NDME in 1994–1995, and the other observations were
identical to those in 1989–1990. In 2000–2001, apparel sector was more capital-
intensive than textile sector in all categories except DME. Urban enterprises were
much more capital-intensive than rural ones. Capital intensity was much higher
during 2000–2001 than that during 1994–1995 in all categories.

While textile sector was less labour-productive than apparel sector in OAME, the
reverse holds true for NDME, during 1984–1985, when urban enterprises and
NDME were more labour-productive than, respectively, rural enterprises and

OAME. This had increased in most categories from 1984–1985 to 1989–1990. Except for the fact that apparel sector was more labour-productive than textile sector in most cases, relative positions remain the same as 1984–1985. In 1994–1995, labour productivity was less in apparel sector than in textile sector in all categories except urban OAME and DME.

Urban enterprises, DME and NDME were more labour-productive than rural enterprises, NDME and OAME, respectively. While labour productivity grew in most of textile sector between 1994–1995 and 2000–2001, with an exception of OAME, apparel sector was more labour-productive than textile sector.

To highlight the findings of this section, two observations are to be mentioned at this point. Firstly, urban enterprises have been performing better than rural enterprises in most subsectors and measures in the unorganized textile sector. This reiterates the dominant problem of rural-urban divide even in this section of the economy. Secondly, DME have performed better than NDME, which have performed better than OAME in this sector. This supports the argument that smaller firms may not be in a position to perform better than larger ones, highlighting the issue of encouraging the relatively susceptible segments of the industry, so as to provide a level-playing field. Having examined the organized and unorganized segments of Indian textile sector, which form its supply side, it is useful to look at some aspects of domestic demand for textiles and clothing. An attempt is made to do this in the next section.

3.4   Domestic Consumption of Textiles in India

Household textile demand holds immense significance in the Indian economy. Given India's population, and more importantly its exploding growth rate, textiles, as a part of the subsistence trio (food, clothing and shelter), are poised to be among the key factors of demand. Tables 3.9 and 3.10 illustrate that the share of clothing in the total expenditure of an average Indian household is around 6–7% in the recent years.

The share of textiles and clothing in total expenditure could be an indicator of development for the countries, because the more the households in a country relatively spend for clothing, the more developed and comfortable they are with their other basic necessities, especially, food. Thus, there seems to be some scope of increasing the per capita demand for clothing, which could show up as an increase in the share of clothing in the total expenditure. In urban households, the share of clothing in non-food expenditure has been much lower than in rural households. This is partly because the basket of non-food commodities (both goods and services) is bigger in urban areas, hence rendering the share of clothing relatively low. However, these shares have been slowly falling in both rural and urban areas.

Further, sickness of various textile mills in the past has been largely attributed to lack of demand in the country by various studies (e.g., see Goswami 1985 and 1990; Murty and Sukumari 1991). Though most of these studies were based on the data

Table 3.9 Trends in per capita consumption expenditures and shares on clothing in rural India (current prices)

Per capita	1989–1990	1993–1994	1999–2000	2000–2001	2001–2002	2002–2003	2003–2004	2004–2005
Expenditure on								
Clothing (Rs.)	10.52	21.20	33.28	35.94	35.33	37.68	38.58	39.80
Non-food	57.28	108.30	197.36	216.34	221.92	239.21	255.68	260.1
Total (Rs.)	158.10	286.10	486.16	494.90	498.27	531.49	555.55	616.57
Clothing's share	0.18	0.20	0.17	0.17	0.16	0.16	0.15	0.15
In non-food								
Clothing's share	0.07	0.07	0.07	0.07	0.07	0.07	0.07	0.07
Total								

Source: Author's calculations from NSSO (2005)

Table 3.10 Trends in per capita consumption expenditures and share of clothing in urban India (current prices)

Per capita	1989–1990	1993–1994	1999–2000	2000–2001	2001–2002	2002–2003	2003–2004	2004–2005
Expenditure on								
Clothing (Rs.)	15.00	32.70	51.76	58.16	57.81	60.83	60.08	62.48
Non-food	110.18	214.00	444.08	514.01	530.48	582.18	593.56	619.74
Total (Rs.)	249.92	464.30	854.92	914.57	932.79	1011.97	1022.68	1104.84
Clothing's share	0.14	0.15	0.12	0.11	0.11	0.10	0.10	0.10
In non-food								
Clothing's share	0.06	0.07	0.06	0.06	0.06	0.06	0.06	0.06

Source: Author's calculations from NSSO (2005)

and scenario till the late 1980s, a demand constraint could be expected to have persistent evidently in the textile sector, at least till 2005, when the MFA quotas were phased out, leading to a boom in the external demand sector. Thus, demand for clothing seems to have two dimensions relevant for a country's development: its own intrinsic value as an indicator of development and its implications for the supply side and hence the employment aspects.

Table 3.11 shows that the aggregate household purchases of textiles have grown over years, though the per capita purchases have either been stagnant or have fallen, unlike the exports, which have been increasing for decades, despite the quota system. The domestic demand trends are not in line with the trends in domestic production, as illustrated. Hence, there is clearly a domestic demand constraint for textiles in India.

Table 3.11 Indian textile and apparel sector: trends in growth of supply and demand

Period	Aggregate household purchases	Per capita household purchases	Exports	Supply (production)
1975–1980	3.519	0.991	3.877	6.35
1980–1985	4.742	2.225	0.402	4.841
1986–1994	0.875	−1.08	14.478	10.518
1995–2000	3.026	1.129	19.045	5.033
2000–2005	4.001	2.028	10.205	8.9

Source: Author's calculations from different yearbooks of Annual Survey of Industries, Compendium of Textile Statistics, Directorate General of Foreign Trade and consumer's purchases in textiles

The demand constraints are attributed to the excise structure that is highly biased towards cotton and other natural fibres as well as the textile commodities that are manufactured by relatively less efficient ways, such as without power and steam. Table 3.13 shows the excise structure, over years in different textile fibres, while Tables 3.14 and 3.15 show the same in different yarns and fabrics, respectively.

Before an examination of the figures in these tables, it is imperative to note a few things. Firstly, natural fibres, hank yarn (plain reel and cross reel up to 25 s), all fabrics processed without aid of power and steam and products of factories owned by/registered to National Handloom Development Corporation, State Government Handloom Development Corporations and Khadi and Village Industries Commission have no excise duty to begin with. Secondly, since 1995–1996, a provision was made in the budget to make a part of excise duty in lieu of sales tax for all fabrics, and hence the figures from this year are a bit higher than what they effectively are, in comparison with those for the previous years. Thirdly, handloom cotton fabrics and those processed by independent power-processors approved by government have an excise duty that is 40% of mill and power loom sector.

Woollen fabrics made of shoddy yarn are exempted up to the value of Rs. 60/sq. metres till 1992–1993 and Rs. 100/sq. metres since 1993–1994. Hank yarn exemption was withdrawn from 2002–2003, but the exemption to hank yarns of coarse counts up to 2 s (English count, i.e. number of 840 yards of yarn in one pound) using condenser card machines is maintained. Since 2004–2005, duties applicable with centralized value-added taxes for natural fibre yarns and all fabrics (Table 3.12).

Considering the fact that the recent figures for excise duties consist of what was sales tax before as well, it can be observed that there is a falling trend in almost all commodity groups. Another inference is that the excise structure is now much simpler than it was before. For example, while it was different for each type of staple fibre before, it is same for all of the synthetic stable fibres in the recent years. Filament yarns in general and polyester in particular are the commodity groups for which the excise duties appear to be the highest.

For simplicity, we have not shown the excise structure of the intermediates involved in the production of synthetics. For most of them, it has remained static at around 15–18% for the past ten years. Thus, it is very clear that the excise structure

Table 3.12 Trends in excise structure of various textile staple fibres in India from 1992 to 2005

Year	Acrylic, viscose	Polyester	Nylon	Acetate	Polypropylene
1992–1993	15.6	13.65	59.15	15.6	17.87
1993–1994	14.95	12.65	14.95	14.95	17.25
1994–1995	23	23	23	23	23
1995–1996	23	23	23	23	23
1996–1997	23	23	23	23	23
1997–1998	20.7	20.7	20.7	20.7	20.7
1998–1999	20.7	20.7	20.7	20.7	20.7
1999–2000	18.4	18.4	18.4	18.4	18.4
2000–2001	18.4	18.4	18.4	18.4	18.4
2001–2002	18.4	18.4	18.4	18.4	18.4
2002–2003	18.4	18.4	18.4	18.4	18.4
2003–2004	18.4	18.4	18.4	18.4	18.4
2004–2005	16.32	16.32	16.32	16.32	16.32

Source: Compendium of textile statistics (annual books published by the Office of Textile Commissioner, Ministry of Textiles, Government of India for the years from 1994 to 2005)

is still highly biased towards the natural fibres, though this has been reduced to a large extent. Further, less efficient ways of manufacturing such as those that do not use power and steam pay less excise duties, leading to higher relative marginal costs of production for the more efficient manufacturers. This kind of differentiation is removed only in the case of woollen fabrics, as noted in the Table 3.14.

A recent exercise on demand estimation, using a dynamic almost ideal demand system, performed for a monthly household-level survey data on textile purchases from 1994 to 2003, done by the author,[19] shows that the cross-price elasticities among the 12 major commodity groups within textiles are negligible compared to the own-price elasticities, which are very high for the synthetic and blended textiles and low for cotton textiles. These findings are in line with the older studies on textile demand, showing that not much has changes in the textile consumption pattern in India over years. This is summarized in Table 3.15, where the own-price elasticities and expenditure elasticities are shown in bold font. It is evident that the cross-price elasticities are negligible compared to these. Further, own-price elasticities are strikingly higher in synthetics than in cotton and wool.

All these observations, put together, point towards two major facts. First one is the biased nature of excise structure that has kept not only synthetic/blended textiles more expensive than they should have been but also has encouraged the less efficient means of production, albeit for developmental purposes such as equity. Second one is that a reduction of this bias by lowering the excise on synthetics/blended textiles and more efficient means of production would not cause a fall in demand for

---

[19] Details of this model, not shown here for simplicity and space constraint, are available on request from the author (Narayanan, 2007).

Table 3.13 Trends in excise structure of various textile yarns based on filaments and staple fibres in India from 1992 to 2005 (in percentage ad valorem)

YYear ear	Hank yarn	Cone yarn	Polyester viscose	Polyester cotton	Polyester wool	Polyester filament	Nylon filament	Viscose filament	Wool yarn
1992–1993	0.39–2.60	0.35–9.75	15.6	7.8	15.6	80.6	25–71.5	5.2–19.5	0
1993–1994	0.23–2.30	0.58–9.78	16.1	8.05	16.1	69	26.5–57.5	5.18–19.55	0
1994–1995	3.45	5.75	23	23	23	69	23–34.5	11.5–17.25	11.5
1995–1996	3.45	5.75	23	23	23	57.5	23–34.5	11.5–17.25	11.5
1996–1997	3.45	5.75	23	23	23	46	23–34.5	11.5–23	11.5
1997–1998	3.45	5.75	20.7	20.7	20.7	34.5	20.7–34.5	9.2–20.7	9.2
1998–1999	3.45	5.75	20.7	20.7	20.7	34.5	20.7–34.5	9.2–20.7	9.2
1999–2000	0	9.2	18.4	18.4	18.4	34.5	27.6	18.4	9.2
2000–2001	0	9.2	18.4	18.4	18.4	36.8	18.4	18.4	9.2
2001–2002	0	9.2	18.4	18.4	18.4	36.8	18.4	18.4	18.4
2002–2003	0–9.20	9.2	18.4	18.4	18.4	36.8	18.4	18.4	18.4
2003–2004	0–9.20	9.2	13.8	13.8	13.8	27.6	13.8	13.8	13.8
2004–2005	0–9.2	9.2	8.16	8.16	8.16	24.48	16.32	16.32	8.16

Source: Compendium of textile statistics (annual books published by the Office of Textile Commissioner, Ministry of Textiles, Government of India for the years from 1994 to 2005)

Table 3.14 Trends in excise structure of various textile fabrics in India from 1992 to 2005

Year	Cotton fabrics	Blended/synthetic fabrics	Woollen fabrics (1)	Woollen fabrics (2)	Woollen fabrics (3)
1992–1993	0.2–2.5 + 20% Of value > Rs. 40/ sq. mtrs.	0.5–20%	2.0–9.0	7.1–14.4	10.86–18.00
1993–1994	0.2–2.5 + 20% Of value > Rs. 40/ sq. mtrs.	0.5–20%	2.0–9.4	7.95–15.50	10.75–18.80
1994–1995	10	10–20%	0–16.50	16.5	16.50–22.25
1995–1996	5–10%	10–20%	22.25	22.25	22.25
1996–1997	10–20%	20	22.25	22.25	22.25
1997–1998	10–20%	20	22.25	22.25	22.25
1998–1999	10–20%	20	22.25	22.25	22.25
1999–2000	13–16	16	21	21	21
2000–2001	16	16	21	21	21
2001–2002	16	16	16	16	16
2002–2003	12	12	12	12	12
2003–2004	10	10	10	10	10
2004–2005	4.08	8.16	8.16	8.16	8.16

Source: Compendium of textile statistics (annual books for the years from 1994 to 2005)
Note 1: The units are percentage ad valorem for all except woollen fabrics, for which the units are rupees per sq. mtr., unless otherwise mentioned
Note 2: 1 – Manufactured by independent processors; 2 – manufactured by decentralized sector and processed by mills; 3 – manufactured and processed by composite mills

the conventional textiles, as the cross-price elasticities hardly play a role in the scene. Further, such a reduction would enhance the demand for all non-cotton commodity groups, without affecting the demand for cotton and other conventional commodity groups.

Given the above description, it is quite understandable that a cut in excise duties of synthetic and blended textiles will be beneficial to the Indian textile sector, as a whole. While presenting the Union Budget for the year 2006–2007, India's Finance Minister has probably had these issues in mind while reducing the excise duty of man-made and blended fibres from 16% to 8%. This is, indeed, a welcome step. While this analysis has focussed only on domestic demand, this also has implications also for India's competitiveness vis-à-vis the other countries in the textile sector, in international trade. With reduced protection, Indian industries are likely to become more competitive, and some raw material inputs are likely to become cheaper due to lowered duties.

Thus, it may be said with a reasonable degree of confidence that the Indian textile sector is going to get benefited immensely because of such steps of tariff and tax reduction. The major point emphasized in this section, which is less obvious, is that a cut in duties will not affect the conventional textiles sector, owing to the low cross-price elasticities between the textile commodity groups. This is not only essential

Table 3.15 Elasticities of various textile commodity groups to their prices and textile expenditure

Elasticity of:	Acrylic	Viscose	Cotton	Cotton-viscose	Nylon	Polyester	Polyester-cotton	silk	Polyester-viscose	Polyester-wool	Wool
With Respect to:											
Acrylic	**-0.851**	0.008	0.045	0.013	0.073	-0.070	-0.036	-0.001	-0.109	-0.033	-0.021
Viscose	0.010	**-0.920**	0.035	0.025	0.134	-0.056	0.046	-0.024	-0.031	-0.002	-0.027
Cotton	0.007	-0.020	**-0.667**	-0.024	-0.054	0.042	-0.323	0.002	0.034	0.036	0.021
Cotton-viscose	0.006	0.010	-0.010	**-0.876**	-0.099	-0.001	-0.017	-0.001	-0.001	0.003	-0.012
Nylon	0.012	0.023	-0.010	-0.037	**-1.334**	0.009	-0.036	0.009	0.046	0.036	-0.0003
Polyester	-0.061	-0.053	0.117	0.001	0.054	**-0.948**	0.188	-0.019	-0.043	-0.032	-0.055
Polyester-cotton	-0.022	0.055	-0.340	-0.023	-0.157	0.198	**-0.906**	0.015	-0.012	0.036	-0.026
Silk	-0.001	-0.067	0.149	0.005	0.158	-0.054	0.025	**-0.936**	0.011	-0.133	-0.089
Polyester-viscose	-0.043	-0.014	0.040	0.001	0.107	-0.020	-0.011	0.001	**-0.688**	-0.042	-0.040
Polyester-wool	-0.021	-0.004	0.044	0.004	0.126	-0.024	0.021	-0.033	-0.066	**-0.730**	-0.045
Wool	-0.035	-0.049	0.110	-0.045	-0.001	-0.109	-0.059	-0.057	-0.179	-0.126	**-0.713**
Textile expenditure	**1.000**	**1.039**	**0.487**	**0.982**	**0.941**	**1.018**	**1.129**	**1.037**	**1.005**	**0.981**	**1.032**

Source: Author's calculations from household-level data from textiles committee (Narayanan 2007a,b)

Note: Values in bold are all significant at 99% confidence interval

for the well-being and better performance of the sectors, per se, but also the standards of living of the masses, in terms of textile consumption. It should be highlighted that consumption of textiles itself is as much a measure of development as is the consumption of food. Hence, enhancing textile consumption should be an inherent feature of developmental policies. In addition, enhanced textile demand would benefit the supply side as well, which is immensely significant for development of the economy in general.

3.5   Conclusions

With an objective of analysing the structure of India's textile sector from both supply and demand perspectives, this chapter has considered the performance and employment in organized and unorganized sectors, the fiscal and tariff policies and their impacts on domestic consumption of textiles and clothing in India.

Examining the organized textile and apparel sector, it is seen that employment has been stagnant, while capital and output have been increasing, till 2000–2001, after which employment has started increasing as well. Apparel sector has expanding tremendously in terms of output, capital and employment, despite a much lower increase in the number of factories than in textile sector, indicating a structural change, in terms of huge investment and increase in scales of operation, since its dereservation from SSI sector in 2000. Better prospects of employment are possible in apparel sector in future, though it should be enhanced in textile sector as well, by promoting huge investments. Even in the unorganized sector, smaller firms are worse off than the bigger ones, in terms of various productivity measures. Hence, even small firms could be encouraged to grow bigger by investing more while preserving their merits in being small, especially, flexible and customized production possibilities.

Investment could be encouraged by better credit disbursement policies. In this connection, it should be noted, however, that the credit disbursement through the Technology Upgradation Fund Scheme (TUFS) scheme, as a fraction of credits applied for, has been decent enough as shown in Table 3.16.[20] A glance at the figures in this table would suggest that the disbursement of credit has been fairly good especially in the case of the agencies that are meant for promoting the SSIs (such as Small Industries Development Bank of India, SIDBI), with an application-rejection rate of less than 2% and credit disbursement rate of around 70%, though the figures are less impressive for the agencies that lend to all industries (such as Industrial Credit and Investment Corporation of India: ICICI, Industrial Development Bank of India: IDBI and Export Import Bank). To the extent that SSIs are more dependent on the sources of credit such as TUFS than the other industries, these figures show that the credit disbursement is not a major issue. In fact, the same can be said for the other industries too, though not to the extent that it can be said in the case of SSIs.

---

[20] See Narayanan (2005) for more details in this regard.

Table 3.16 Credit applications that were received and disbursed under TUFS (2004–2005)

Nodal agencies	Credit applications received			Credits disbursed			No. of applications rejected
	No. of applications	Project cost	Amount of loan required	No. of applications	Project dost	Amount sanctioned	
Agencies that lend to all industries	1290	23031.07	12237.79	950(73.64)	14224.00 (61.68)	6682.58 (55.00)	118 (9.15)
Agencies that lend only to SSI	2379	2498.38	1480.32	1930(81.13)	1778.29 (71.18)	1006.88 (68.04)	44 (1.85)
Total	3669	25529.45	13718.11	2880(78.50)	16002.29 (62.26)	7689.46 (56.04)	162 (4.42)

Source: Author's calculations from a report on 'Progress of TUFS as on 30.11.2004,' by the Office of the Textile Commissioner, Mumbai. *Note*: Figures corresponding to costs/amount in the table are in crores of rupees (1 crore = 10 million), and those in brackets are percentage of the corresponding total

Thus, the reasons for the low investment may be the lack of awareness among the entrepreneurs about these schemes, and hence the government should take steps to promote such useful schemes.

As for the unorganized textile sector, employment has been increasing despite fall in capital and output, an issue that is in striking contrast with that in organized textile sector. In late nineties till 2001, capital productivity has declined in this sector, more so in urban than in rural areas. Capital intensity was much higher during 2000–2001 than that during 1994–1995 in all categories. While labour productivity grew in most of textile sector between 1994–1995 and 2000–2001, with an exception of OAME, apparel sector was more labour-productive than textile sector. Enterprises in rural areas were more capital-productive, less capital-intensive and less labour-productive than those in urban areas. Apparel sector was less capital-productive, more capital-intensive (except in DME) and more labour-productive than in textile sector. These trends varied across enterprise types as well. A major observation from the analysis of unorganized textile sector is that there has been a division between various segments within the textile sector, in terms of performance.

The analysis of household demand has shown that the per capita textile purchases have been declining in real terms during the past few years. The excise and customs duties on the man-made fibre textiles have been a barrier in increasing their purchases due to the fact that they are reflected in their prices and that the demand for these products is highly own-price elastic. Given the fact that the cross-price elasticity between cotton and these fibres is negligible compared to the own-price elasticities, rise in demand for textiles in India without a fall in the demand for the conventional textiles could be ensured by fall in prices of man-made fibre textiles, which is possible only by cut in excise duties and customs of these products, as has been done during the recent years. This appears to be a significant step to foster development in the country, from the viewpoints of supply side as well as demand side.

Chapter 4
Environmental Issues

Environmental hazards associated with industrial activities have been predominant for the past few centuries all over the world. Before the advent of the industrial revolution, such hazards were almost unheard of. Most products were made from organic and natural inputs, causing less or no pollution. Today, pollution has become so rampant that the health of millions of people is at risk due to polluted water, air and land. Several modern techniques to control pollution have often resulted in process improvements, but quite a few of them are costly, and some of them even cause new unanticipated problems.[1] Therefore, it is high time industrialists, and academicians alike consider natural solutions from the ancient and medieval world, in terms of reducing pollution in a cost-effective way. Textile sector is a suitable example in this regard, since it had evolved and matured thousands of years ago but has been facing pollution issues in countries like India today.

Historically, natural dyes were used to colour clothing or other textiles, and by the mid-1800s, chemists began producing synthetic substitutes for them. By the early part of this century, only a small percentage of textile dyes were extracted from plants. Lately, there has been increasing interest in natural dyes, as the public becomes aware of ecological and environmental problems related to the use of synthetic dyes. Use of natural dyes cuts down significantly on the amount of toxic effluent resulting from the synthetic dye process.

Natural dyes generally require a mordant, which are metallic salts of aluminium, iron, chromium, copper and others, for ensuring the reasonable fastness of the colour to sunlight and washing. Customers who have become accustomed to the dazzling colours and wash and light fastness of synthetic dyes are hard to convince, as only a few of the natural dyes have good all-round fastness. Most of the natural dyes produce comparatively dull shades. But there are chemicals and methods available now, which can enhance the colour and fastness of the natural dyes. This raises

---

[1] For example, the process of effluent treatment does address the problem of toxicity in water, but it results in accumulation of sludge, which can cause land degradation.

© The Author(s) 2018
B.N. Gopalakrishnan, *Economic and Environmental Policy Issues in Indian Textile and Apparel Industries*, SpringerBriefs in Environmental Science, DOI 10.1007/978-3-319-62344-3_4

the question of whether the natural dyes really reduce emissions or not, since some additional inorganic chemicals are required for them.

The objectives of this study are to analyse the environmental emissions that could be reduced by the introduction of natural dyes in the Indian textile industry by the year 2020 under various scenarios, to prioritize the dyes for different fibres based on their emission levels, costs and fastness and to propose some policy implications for emission reduction using the natural dyes in the Indian textile dyeing sector. In this study, the following sectors and dyes have been chosen for analysis of emissions.

Sectors: Natural fibres: cotton, wool and silk
                  Synthetic fibres: polyester and nylon
Dyes: Natural dyes: fenugreek, golden dock, gallnut, dolu, gum arabic,
                  indigo, shellac, kamala, madder, and pomegranate
                  Synthetic dyes: acid dyes, reactive dyes, sulphur dyes, azo dyes,
                  disperse dyes, direct dyes

To verify the claim that natural dyes reduce pollution in textile processing industries, a multi-scenario analysis of adaptation of natural dyes has been done under different rates of adaptation to natural dyes in different subsectors till 2020. The overall results confirm that natural dyes do cause significant reductions in the total dissolved solids (TDS) by 2020. Based on analytical hierarchical programming (AHP) analysis to rank dyes, natural dyes were found to be better than other dyes in many cases, though there were quite a few cases in which synthetic dyes were better. Encouragement of research in the application of natural dyes, especially on synthetic fibres, implementation of better emission standards and planning for adaptation of natural dyes at an optimal pace are the policy implications.

4.1 Literature Review

A wide range of literature is available in the context of various technological research and developments taking place in the field of the natural dyes. But very few published papers are available in the area of emission analysis of the dyes as a whole. There is no significant work done in the comparison of emission levels due to natural dyes and those due to synthetic dyes.

G.R. Pophalia (2003) analyses the emission levels of synthetic dyes in India to compare them with those resulting from the application of CETP technology in the dyeing houses. This data was useful in checking whether the theoretically calculated emission levels of different dyes lie within the practical range. www.alpsind.com was the source of data for the dyeing procedures for different natural dyes, which are noted throughout this paper using their trademark names of Alps Industries, Ghaziabad. Many other websites cited in the references at the end have been useful for the data regarding the dyeing procedures of synthetic dyes, production figures and the standards of emission levels.

T. Bechtold (2003) discusses the various merits and demerits of natural dyes in the current industrial situations in a technical manner. Dheeraj (2003) has tested the eco-friendliness of natural dyes using the values of effluent concentrations obtained from informal sector-dyers in Rajasthan. Agarwal (2003) analyses the compatibility of natural dyes to the modern dyeing houses. Holme (2003) lists the recent developments in the textile dyes, of which some facts and figures about natural dyes sound significant. Rubina et al. (2002) have analysed the biochemical features of the natural dye effluents such as toxicity and their biodegradability. Patra A.K. (2002) has studied the performance of some natural dyes on the cotton fabrics in terms of fastness and depth. Teli (2000) has given an overall account of natural dyes – their classification, technology and future prospects. Paul (1996) has a description of extraction and application of various natural dyes on different fibres.

Nousianen (1997) has a detailed explanation of the natural dyes. According to this author, environmental awareness is setting tough demands for textile dyeing technology, at the same time as keen competition in the world of fashion textiles requires quick responses, quality and low production costs. Demand is increasing for eco-textiles, certified with various official labels defining colour fastness and the maximum concentrations of toxic or harmful impurities. The standard methodologies that have been developed for certification are also ensuring the use of best available technology (BAT) in production processes. Although natural dyes can serve as models, there is no way that the annual consumption of 0.5 million tonnes of textile dyes can be met by natural dyes.

According to Parham (1994), new technology in application of natural dyes allows a wide range of shades with good fastness properties, no salt use, elimination of heavy metals in processing, safe effluent and rational methods for sourcing raw materials with pos. effects to the environment. Glover (1995) argues that synthetic dyes are more user- and environmentally friendly than natural dyes. The arguments put forward are those related to the excessive use of supporting chemicals, possible over-intensification of agriculture and low quality of natural dyes.

Kharbhade (1988) has described the identification of natural dyes in historic textiles from the mid-nineteenth century. Seventy samples taken from museum textiles were compared with reference materials prepared in the lab as well as with CP major dye components of natural dyes. One yellow, two brown, one blue and two red natural dyes were identified in different primary and secondary coloured samples. It is also seen from the results that mixtures of two dyes have been used to obtain desired shades. Bahl (1988) has developed a method for mordant dyeing of silk with cutch natural dyes in the presence of different mordants (alum, chrome, CuSO4, FeSO4) for a wide variety of colours. The dyed fabrics exhibited fair-to-good wash fastness and light fastness. Moses (2003) has applied the waste grape skin for the application on protein textiles such as silk and wool in combination with other similar natural resources like pomegranate rind, orange rind and annatto seed and has observed that the results such as colours obtained, fastness towards rubbing, light, heat and washing are very much convincing.

In Auslander et al. (1977), naturally occurring dyes were discussed in a form of a monograph with screening studies on their stability. Duff et al. (1977a) have assessed wash fastness of natural dyes on wool by the tests conducted under standard conditions. A number of the dyes underwent marked changes in hue on washing that was due to the effect of small amounts of alkali in washing mixtures. In Duff et al. (1977b), the colour changes accompanying light fading were followed visually and instrumentally and characterized in terms of light-fastness grades, grey-scale differences and colour differences in CIELAB colour space and in Munsell colour coordinates. The light-fastness grades were similar to those previously reported for daylight exposures. The various methods of expressing colour change gave detailed and quantitative measures of the effects of light on the colour of natural dyes; this type of information could be of value in predicting the original colour of ancient textiles.

Kirtikar (1947) and Gulati (1949) have analysed the Indian indigenous natural dyes in terms of their eye appeal, colour value and fastness and found them to be less valuable than the synthetic dyes. An interesting study by Furry (1945) has shown that the cotton fabrics dyed with natural dyes such as osage orange and quercitron showed excellent mildew resistance that persisted even after 6 weeks weathering.

From the literature, it emerges that the natural dyes are, in general, inferior to synthetic dyes in colour value or eye appeal and colour fastness. However, there has been a significant progress in the research of enhancing the dyeing characteristics of the natural dyes for the past five decades, and today, there are many natural dyes with colour value and fastness comparable to those of synthetic dyes. There is a debate among various technologists, researchers and scientists regarding the utility of natural dyes. While some of them claim that natural dyes can be widely used as 'better substitutes' for the synthetic dyes in terms of emissions and cost, others claim that the colour value is not as good in the case of natural dyes as in the synthetic dyes. A betterment of colour value of natural dyes would involve use of some chemicals, which in turn, would cause pollution again.

Some researchers have even expressed their doubt over the validity of the widely spoken idea that natural dyes essentially reduce the pollution levels. A small proportion of the scientists predict that substitution of the synthetic dyes with the natural dye would cause serious problems of environmental concern in terms of over-intensification of agricultural land, deforestation, etc. Hence, to end up with a meaningful decision, a balanced multi-scenario approach that considers all relevant aspects of natural dyes and synthetic dyes in a macro-level, for the entire sector in the country, with proper projections about the future with different assumptions on the adaptation rates of natural dyes, is necessary.

4.2   Methodology

4.2.1   Steps Involved for the Multi-scenario Analysis

Step 1: the Indian production figures of dyed yarn and cloth made of cotton, silk, wool, polyester and nylon are projected for 2005, 2015, 2025 and 2030, based on the average annual growth rate calculated from the data available for the years 1985–1999.

Step 2: the inorganic chemical composition or total dissolved solids (TDS) of the effluents generated by dyeing different materials for different colours using natural and synthetic dyes is calculated using the procedure explained.

Step 3: the emission levels (kg of inorganic chemicals per kg production of the material) calculated as above are compared with the standards fixed by the government of India.

Step 4: different scenarios are constructed based on different rates of adaptation of the natural fibres and synthetic fibres sectors to the natural dyes.

BAU (business as usual): 100% of the dyes used for all the categories of materials are synthetic throughout the period of analysis, except for silk, in which 70% are natural dyes. For the proposition that natural dyes reduce emissions to be true, BAU should cause maximum emissions.

Scenario 1 (optimistic): the shares of natural dyes and synthetic dyes for different years start at 20% in 2005 and increase by 10% points every 5 years, except for silk for which 70% of natural dyes is assumed for all periods. This scenario would be expected to cause least emissions if natural dyes were 'pollution reducers'.

Scenario 2 (realistic): for cotton, wool and silk, the same percentage of dyes is maintained as in scenario 1; while for synthetic fibres, the rates of 0% for 2005 and a rise of 10% points every 5 years are assumed. Hence, this scenario is realistic given the lower applicability of natural dyes on natural fibres. To claim that natural dyes reduce pollution, scenario 2 should give second least level of emissions, since it has the second fastest adaptation rates for the synthetic fibres and fastest rates for the natural fibres.

Scenario 3 (pessimistic): in this scenario, the assumption is that the natural dyes are not adapted to synthetic fibres throughout the period of analysis and that they are adapted to natural fibres at a pace at which they were assumed to be adapted to synthetic fibres in the scenario 2. Silk is assumed to use natural dyes at 70% throughout the period. Though this scenario is termed as pessimistic one, it should be noted that even this is not as pessimistic as the BAU, in which the natural dyes were assumed to be totally absent in all sectors throughout the period of study. Hence, this scenario should give the second largest level of emissions.

Scenario 4 (no synthetic fibres with natural dyes): natural dyes are adapted by the natural fibres at a pace defined in scenario 1, while synthetic fibres do not at all adapt them. As the natural fibres adapt to natural dyes at higher pace in this scenario than in scenario 3, this scenario should cause less emissions than scenario 3, i.e. third largest level of emissions.

Step 5: based on the emission levels obtained under each scenario, different questions could be raised so as to suggest few policy implications.

4.2.2 Assumptions Involved in the Multi-scenario Analysis

The production of the materials grows at an average annual rate calculated from the data for the years 1985 to 1999. For synthetic dyes, 40% of chemicals added in the recipe are retained in effluent. For both the synthetic and natural dyes, 10% of the chemicals remain in the effluent after proper treatment in ETP. For the natural dyes, 70% of chemicals added in the recipe are retained in effluent. Light, medium and dark shades form equal proportions of the total materials dyed.

Silk is dyed with 50% natural dyes now, and it may not go beyond 70% in the future. The measure of emissions is GPL of TDS (total dissolved solids – only inorganic material), and hence it does not include any organic material in the effluent. Hence, natural dyes are assumed to be non-polluting in nature as they are organic. The rates of adaptation have been designed based on the fact that many synthetic dyes are being banned (e.g. azo dyes) in the international market and many more are anticipated to be banned in the future.

The extent to which the sectors are expected to adapt the natural dyes is based on the expectation of the other dyes to be banned, and there is no concrete set of data available in this regard. Hence, we assume some reasonable adaptation rates, and the inclusion of different scenarios would serve the purpose of including all possible ranges of adaptation.

4.2.3   Steps Involved in the AHP-Based Ranking of the Dyes

AHP is the advanced hierarchical programming technique, which is used to prioritize among various options based on qualitative and/or quantitative criteria.

Step1: the emission levels of different dyes for different shades in the case of different sectors are calculated in terms of GPL (grams of TDS per litre of effluent).

Step2: for each shade, dye and sector, the values of GPL, cost and fastness index (average of light fastness, wash fastness and rubbing fastness) are listed.

Step3: the reciprocals of GPL and cost and the fastness index value are used for the AHP analysis. A matrix is formed for each dye-shade-sector combination, whose rows are the different dyes and columns are 1/GPL, 1/cost and fastness index.

Step4: based on a survey conducted with industrialists, academicians and research students, the weights for each of the criteria are calculated and written in vector form. The matrix obtained in step 3 above is multiplied by this vector.

Step5: the vector resulting from the step 4 above is normalized, and based on the values of this normalized vector, the dyes are ranked. The higher the value, the higher is the rank.

Assumptions Involved in the AHP-Based Ranking of the Dyes
All criteria are quantitative, i.e. possibilities of taking durability (rather than fastness) or pollution/health hazards (rather than emission level in GPL) as a qualitative criteria are ruled out. The three criteria considered here are assumed to be the only ones for the ranking of dyes used for a same type of shade. International prices are taken as the cost criteria.

4.3 Results

4.3.1 Emission Levels of Different Dyes

Our results show that natural dyes are the most preferable for cotton, at least in the context of the environmental emissions for the light and dark shades. This is not so in the case for the medium shade since golden dock is a natural dye requiring many supporting chemicals and sulphur dye causes less TDS in all cases. We also see that natural dyes cause less emission than the synthetic dyes in the case of wool. For silk, natural dyes are least polluting, and the difference in GPLs of TDS for natural and other dyes is significant. For polyester, natural dyes are more polluting than the synthetic disperse dyes. This is to be taken as evidence in favour of the proposition that dyeing of natural dyes, on synthetic fibres, is so complicated and cumbersome a process that it would necessitate the addition of quite a few supporting chemicals, due to which the pollution caused by natural dyes turns out to be higher than that caused by the corresponding synthetic dyes.

In the case of nylon, unlike polyester, the natural dyes are again less polluting than synthetic dyes because of the fact that nylon requires a high concentration of the synthetic dyes and its supporting chemicals for its dyeing that even the natural dyes, despite causing a higher level of pollution than in other sectors, cause a lower level of pollution than the synthetic dyes.

4.3.2 LEAP Results of the Scenario Analysis

LEAP (Long-Range Energy Alternatives Planning) is a software package used to calculate the pollution levels and energy consumption levels of any system based on its various branches and for different years in the future for different scenarios. In our case, we do not include any analysis on energy consumption in this context but focus on environmental aspects alone. This is a major deviation from routine uses of LEAP. The system may consist of different sectors, each of which may, in turn, be comprised of subsectors. Each subsector would have different end uses, and each of which would make use of different pollution-causing elements. In this study, the sectors are the different fibres, the subsectors are the different shades, and the end use is the 'type of dye'. The last level in this hierarchy tree is the name of the particular dye, for which the emission level is fed as input in GPL. This exercise is repeated for each of the scenarios mentioned earlier with proper feeding of the data regarding the adaptation rates for the different years. The data of production in each sector, with its share in each level, is also fed into LEAP.

The average annual growth rates of Table 4.1 were calculated based on the data available for the years 1985 to 1999. A glance at the rates and figures would confirm that polyester would dominate the textile industry, comprising of more than 90% of

Table 4.1 Production of dyed fabrics and yarns in the sectors in billion kgs

Sector	1990	Growth rate (%)	2005	2010	2015	2020
Cotton	1.306	5.51	2.92	3.818	4.992	6.528
Wool	0.024	4.55	0.047	0.058	0.073	0.091
Silk	0.02	11.33	0.1	0.171	0.293	0.5
Polyester	0.21	22.06	4.177	11.316	30.659	83.065
Nylon	0.021	0.33	0.022	0.022	0.023	0.023

Table 4.2 Overall results (LEAP) of emissions of TDS under different scenarios

Scenario		2005	2010	2015	2020
BAU	Total (billion kgs)	0.64	1.56	4.05	10.86
	GPG	0.0881	0.1014	0.1124	0.1204
	GPL	*4.4041*	*5.0699*	*5.6188*	*6.0195*
Scenario 1	Total (billion kgs)	0.5195	1.1215	2.5444	5.8983
	GPG	0.0715	0.0729	0.0706	0.0654
	GPL	*3.5745*	*3.6447*	*3.5299*	*3.2693*
Scenario 2	Total (billion kgs)	0.64	1.55	4.02	10.81
	GPG	0.0881	0.1008	0.1115	0.1198
	GPL `	*4.4041*	*5.0374*	*5.5771*	*5.9918*
Scenario 3	Total (billion kgs)	0.5916	1.3161	3.0708	7.6798
	GPG	0.0814	0.0856	0.0852	0.0851
	GPL	*4.0713*	*4.2773*	*4.2602*	*4.2568*
Scenario 4	Total (billion kgs)	0.5916	1.4134	3.5971	9.4613
	GPG	0.0814	0.0919	0.0998	0.1049
	GPL	*4.0713*	*4.5935*	*4.9905*	*5.2442*

the total production in 2020. This, in turn, would imply that polyester sector, in addition to being one with high pollution intensity, is going to be the major contributor to the pollution because of its size effect. This observation, when linked with another note made in the previous section that the technology of application of natural dyes on polyester is not very advanced, would imply that a concrete policy to encourage the technical research in this area is essential for a significant emission reduction.

Based on Table 4.2, many inferences can be made, such as the following:

1. Maximum possible reductions in the emission of TDS by the penetration of natural dyes (at the most optimistic pace) = E(BAU)-E(Scenario1) = 4.96167 billion kg. E(X) is the emission levels in scenario X, as seen in Table 4.2.
2. Emission reductions by promotion of natural dyes for both synthetics and natural fibres at equal pace (rather than depriving synthetic fibres of natural dyes) = E(scenario4)-E(scenario1) = 3.56296 billion kg (a reduction in GPL of 1.98).
3. Emission reduction of natural dyes is adapted to synthetic fibres to some extent and to natural fibres to a much greater extent = E(BAU)-E(scenario2) = 3.1802 billion kg.

4. Emission reduction of natural dyes is not at all adapted to synthetic fibres and to natural fibres to some extent, as compared to the BAU = E(BAU)-E(scenario4) = 1.3987 billion kg.
5. With relatively realistic assumptions, emission reductions as compared to BAU are 3.1802 billion kg. Hence, the natural dyes cause a clear drastic reduction in emissions of TDS.
6. Emission reductions in the case of equally fast promotion measures of natural dyes for both natural and synthetic dyes, compared to:

   (a) BAU: 4.96167 billion kg (45.69% reduction)
   (b) Scenario 3: 4.9117 billion kg (45.44% reduction)
   (c) Scenario 2: 1.782 billion kg (23.2% reduction)
   (d) Scenario 4: 3.56296 billion kg (37.66% reduction)

7. Even in terms of grams of TDS emitted per gram of dyed material (GPG) and grams per litre of TDS in the effluent (GPL), the reductions are significant. As compared to the different scenarios, the final reduction in GPL by the year 2020 of the scenario 1 is implemented as follows:

   (a) BAU: 2.75 GPL (45.69% reduction)
   (b) Scenario 3: 2.72 GPL (45.44% reduction)
   (c) Scenario 2: 0.987 GPL (23.2% reduction)
   (d) Scenario 4: 1.98 GPL (37.66% reduction)

The basic purpose of the above inferences is to quantify the benefits of adapting to natural dyes in terms of TDS and percentage. More importantly, this exercise aims at precisely disaggregating the consequences of adaptation in different sectors at different rates, which is solved by the definition of scenarios. For example, the reductions by not adapting to natural dyes only in synthetic fibre sectors compared to those by uniformly adapting to all sectors would be as shown in inference 2.

As seen in Table 4.3, the cotton sector has least emissions in three scenarios – scenario 1, scenario 2 and scenario 4. This is because of the fact that the adaptation rates for this sector have not changed in these scenarios. The emission reductions are about 30% by the adaptation of natural dyes, i.e. from 302.53 million kg or 2.317 GPL in BAU to 212.26 million kg or 1.626 GPL in these scenarios. Even in the scenario 3, the reductions are almost 20%, i.e. from 302.53 million kg or 2.317 GPL in BAU to 257.4 million kg or 1.972 GPL in scenario 3. This implies that, by 2020, the cotton sector would reduce its emissions by 20% under pessimistic rates of adaptation of natural dyes and 30% under normal rates of adaptation.

As seen in Table 4.4, the wool sector has least emissions in three scenarios – scenario 1, scenario 2 and scenario 4. This is because of the fact that the adaptation rates for this sector have not changed in these scenarios. Just similar to the cotton sector, by 2020, the wool sector would also reduce its emissions by 20% under pessimistic rates of adaptation of natural dyes and 30% under normal rates of adaptation.

Table 4.3   Emission of TDS in cotton dyeing sector

Scenario	Years →	2005	2010	2015	2020
BAU	Total emissions in million kg	135.32	176.94	231.37	302.53
	GPG	0.046342	0.046344	0.046348	0.046343
	GPL	2.317123	2.317182	2.317408	2.317172
Scenario 1	Total emissions in million kg	119.17	145.27	176.14	212.26
	GPG	0.040812	0.038049	0.035284	0.032515
	GPL	2.040582	1.902436	1.764223	1.625766
Scenario 2	Total emissions in million kg	135.32	166.38	203.75	257.4
	GPG	0.046342	0.043578	0.040815	0.03943
	GPL	2.317123	2.178889	2.040765	1.971507
Scenario 3	Total emissions in million kg	119.17	145.27	176.14	212.26
	GPG	0.040812	0.038049	0.035284	0.032515
	GPL	2.040582	1.902436	1.764223	1.625766
Scenario 4	Total emissions in million kg	119.17	145.27	176.14	212.26
	GPG	0.040812	0.038049	0.035284	0.032515
	GPL	2.040582	1.902436	1.764223	1.625766

Table 4.4   Emission of TDS in wool dyeing sector

Scenario	Years→	2005	2010	2015	2020
BAU	Total emissions in 1000 kgs	2045.73	2556.68	3195.24	3993.3
	GPG	0.043526	0.044081	0.04377	0.043882
	GPL	2.176309	2.204034	2.188521	2.194121
Scenario 1	Total emissions in 1000 kg	1793.52	2084.71	2409.66	2767.11
	GPG	0.03816	0.035943	0.033009	0.030408
	GPL	1.908	1.797164	1.650452	1.52039
Scenario 2	Total emissions in 1000 kg	2045.73	2400.69	2805.34	3384.18
	GPG	0.043526	0.041391	0.08429	0.037189
	GPL	2.176309	2.06956	1.921466	1.85944
Scenario 3	Total emissions in 1000 kg	1793.52	2084.71	2409.66	2767.11
	GPG	0.03816	0.035943	0.033009	0.030408
	GPL	1.908	1.797164	1.650452	1.52039
Scenario 4	Total emissions in 1000 kg	1793.52	2084.71	2409.66	2767.11
	GPG	0.03816	0.035943	0.033009	0.030408
	GPL	1.908	1.797164	1.650452	1.52039

Table 4.5 Emission of TDS in silk dyeing sector

Scenario	Years→	2005	2010	2015	2020
BAU	Total emissions in million kg	0.42	5.86	10.01	17.13
	GPG	0.0342	0.034269	0.034164	0.03426
	GPL	0.4275	0.428363	0.427048	0.42825
Scenario 1	Total emissions in million kg	3.42	5.86	10.01	17.13
	GPG	0.0342	0.034269	0.034164	0.03426
	GPL	0.4275	0.428363	0.427048	0.42825
Scenario 2	Total emissions in million kg	3.42	5.86	10.01	17.13
	GPG	0.0342	0.034269	0.034164	0.03426
	GPL	0.4275	0.428363	0.427048	0.42825
Scenario 3	Total emissions in million kg	3.42	5.86	10.01	17.13
	GPG	0.0342	0.034269	0.034164	0.03426
	GPL	0.4275	0.428363	0.427048	0.42825
Scenario 4	Total emissions in million kg	3.42	5.86	10.01	17.13
	GPG	0.0342	0.034269	0.034164	0.03426
	GPL	0.4275	0.428363	0.427048	0.42825

In the case of the silk dyeing sector, as seen in Table 4.5 above, there is no change in emission level because of the fact that the assumptions on the adaptation rates of natural dyes for the silk sector are the same for all scenarios.

As seen in Table 4.6, the polyester sector has least emissions in scenario 1. The emission reductions are about 45% by the adaptation of natural dyes, i.e. from 10.53 billion kg or 6.338 GPL in BAU to 5.66 billion kg or 3.41 GPL in scenario 1. Even in the scenario 2, the reductions are almost 30%, i.e. from 10.53 billion kg or 6.338 GPL in BAU to 7.44 billion kg or 4.478 GPL in scenario 2. This implies that, by 2020, the polyester sector would reduce its emissions by 30% under modest (as in scenario 2) rates of adaptation of natural dyes and 45% under scenario 1 rates of adaptation. This indicates the significance of development of technology of applying the natural dyes on the polyester sector, arising out of the huge potential of emission reductions in this sector.

Table 4.7 shows that a maximum of 42% reduction of TDS is possible by 2020 in the case of nylon, since the difference between the emissions of BAU and OPTI scenarios is (3562.27–2074.55) million kg or (7.744–4.51) GPL. Considering the scenario 2, the possible reduction is 21%. Thus, the potential to reduce the emissions in the nylon sector by adapting the natural dyes is high (21–42%).

A striking observation from Table 4.8 of standards of TDS concentration and emission levels in different sectors is that in most of the scenarios of almost all sectors, the emission levels are too high to be within the standards. Only in the case of silk, in which the natural dyes are most predominant, the standards are satisfied.

Table 4.6   Emission of TDS in polyester dyeing sector

Scenario	Years→	2005	2010	2015	2020
BAU	Total emissions in billion kg	0.5	1.37	3.8	10.53
	GPG	0.1197	0.121068	0.123944	0.126768
	GPL	5.985157	6.053376	6.197201	6.33841
Scenario 1	Total emissions in billion kg	0.39224	0.96568	2.35347	5.66411
	GPG	0.093905	0.085338	0.076763	0.068189
	GPL	4.695236	4.266879	3.838139	3.409444
Scenario 2	Total emissions in billion kg	0.5	1.37	3.8	10.53
	GPG	0.119703	0.121068	0.123944	0.126768
	GPL	5.985157	6.053376	6.197201	6.33841
Scenario 3	Total emissions in billion kg	0.46387	1.15975	2.87927	7.44484
	GPG	0.06896	0.07743	0.08021	0.0896
	GPL	4.1423	4.2467	4.3465	4.478
Scenario 4	Total emissions in billion kg	0.5	1.37	3.8	10.53
	GPG	0.119703	0.121068	0.123944	0.126768
	GPL	5.985157	6.053376	6.197201	6.33841

Table 4.7   Emission of TDS in nylon dyeing sector

Scenario	Years→	2005	2010	2015	2020
BAU	Total emissions in 1000 kg	3390.51	3446.82	3504.07	3562.27
	GPG	0.15411	0.1567	0.1524	0.1549
	GPL	7.7057	7.8337	7.6175	7.7441
Scenario 1	Total emissions in 1000 kg	2824.11	2583.12	2333.34	2074.55
	GPG	0.1284	0.1174	0.1015	0.0902
	GPL	6.4184	5.8707	5.0725	4.5099
Scenario 2	Total emissions in 1000 kg	3390.51	3446.82	3504.07	3562.27
	GPG	0.15411	0.1567	0.1523	0.1549
	GPL	7.7057	7.8337	7.6175	7.7441
Scenario 3	Total emissions in 1000 kg	3390.51	3158.92	2918.7	2818.41
	GPG	0.15411	0.1436	0.1269	0.1225
	GPL	7.7057	7.1794	6.345	6.127
Scenario 4	Total emissions in 1000 kg	3390.51	3446.82	3504.07	3562.27
	GPG	0.1541	0.1567	0.1524	0.1549
	GPL	7.7057	7.8337	7.6175	7.7441

Table 4.8 Indian standards of TDS concentration in effluents

STDS:	Inland surface water		Public sewers		Land for irrigation		Marine coastal areas	
M:L	1:20	1:80	1:20	1:80	1:20	1:80	1:20	1:80
GPG	0.0113	0.0452	0.0279	0.1116	0.0137	0.0548	0.0133	0.0532
GPL	0.5652		1.3952		0.6854		0.6654	

In the cotton and wool sectors, the emission levels are relatively lower than others because of their better adaptability to natural dyes. Polyester and nylon cause the highest GPL emissions of TDS, because of their non-compatibility with natural dyes, being synthetic fibres.

Most significantly, cotton and wool sectors can be, on an average, developed as the sectors with emission levels conforming to the standards under the optimistic scenario (scenario 1). All these observations, in addition to the previous inferences from the overall and sector-wise results, lead us to conclude that natural dyes not only reduce the emissions significantly but also to an extent that at least the natural fibre sectors can conform to the Indian standards for the effluents. As for the synthetic fibres, it would depend on the technological developments that could improve the compatibility of natural dyes with them.

Ranking of the Dyes Based on AHP

For the different sectors, the dyes have been ranked based on three criteria for each shade – cost, colour fastness and emission level (Table 4.10). The international prices and fastness indices of various dyes have been tabulated in Table 4.9.

The inclusion of cost and fastness as the criteria for ranking the dyes, in addition to the emission levels, has changed the order of ranking. The natural dye has moved to second and third place, respectively, for dark and medium shades. This may be due to the fact that the cost of golden dock and gallnut is not low enough to compensate their low fastness value (Table 4.11).

Table 4.12 shows that only golden dock natural dye has been rejected for the first place due to its low fastness, relatively high cost and not low enough emissions. The other natural dyes have retained their first place despite the inclusion of the new criteria.

Table 4.13 would again support the proposition that natural dyes are better than synthetic dyes even when cost and fastness are included as criteria but for the fact that madder natural dye has become least preferred after the inclusion of these criteria. An important note to be made here is that, though azo dye causes low emission of TDS, the content of this effluent is highly toxic and hence azo dye has been banned widely. In this case, a measure of toxicity such as LC_{50} should have been included in the analysis. However, this would matter much only in the case of azo dyes, and since the objective is more towards checking for the feasibility of natural dyes, it can be safely concluded that, for the medium shade, madder natural dye is clearly not the preferable one and is inferior at least to the reactive dye.

Table 4.9   Costs and fastness indices of different dyes

S. no.	Dye	Cost (US$/Lb)	Fastness index
1	Acid dye	25	5
2	Azo dye	20	5
3	Direct dye	20	5
4	Disperse dye	26	5
5	Reactive dye	30	5
6	Sulphur dye	19	5
7	AmberM	43	4
8	Kamala	8.25	3
9	Gum arabic	18	4
10	Fenugreek	43	4
11	Dolu	27.5	4
12	Madder	25	4
13	Indigo	28	4
14	Pomegranate	19	4
15	Shellac	14	4
16	Sahara	14	3

Table 4.10   Ranking of the dyes for cotton

Rank	Light shade	Medium shade	Dark shade
1	Fenugreek*	Sulphur	Sulphur
2	Sulphur	Acid	AmberM*
3	Direct	Sahara*	Reactive

*natural dyes

Table 4.11   Ranking of dyes for wool

Rank	Light shade	Medium shade	Dark shade
1	Dolu*	Acid	Gum arabic*
2	Acid	Sahara*	Acid

*natural dyes

Table 4.12   Ranking of dyes for silk

Rank	Light shade	Medium shade	Dark shade
1	Pomegranate*	Azo	Shellac*
2	Azo	Reactive	Azo
3	Reactive	Madder*	Reactive

*natural dyes

Table 4.13   Ranking of dyes for polyester

Rank	Light shade	Medium shade	Dark shade
1	Disperse	Disperse	Shellac*
2	Kamala*	Indigo*	Disperse

*natural dyes

Table 4.14 Ranking of dyes for nylon

Rank	Light shade	Medium shade	Dark shade
1	Dolu*	Disperse	Gum Arabic*
2	Disperse	Sahara*	Disperse

*natural dyes

In the case of polyester, the natural dye shellac has become a better dye than disperse dye with the inclusion of additional criteria, because of its sufficiently low cost and/or sufficiently high fastness and/or sufficiently low emission. For light and medium shades, natural dyes retain their last rank Table 4.14. This reiterates the importance of encouragement of innovations in the area of application of natural dyes to synthetic fibres, especially, polyester.

As in the most other cases, the natural dye golden dock has been found to be less preferred to synthetic dyes with the inclusion of other criteria for nylon. For light and medium shades, natural dyes still dominate the first place.

4.4 Conclusions

Adaptation to natural dyes would cause a significant reduction 20% to 50% in the emission of TDS by the year 2020 depending on the rate of adaptation. A reasonably fast (OPTI scenario) adaptation would bring the emission levels for the natural fibres (cotton, wool and silk) well within the Indian standards. For reducing the emissions within the standard limits in the case of synthetic fibres (polyester and silk), a combination of fast adaptation and encouragement of research in the application of natural dyes on the synthetic fibres is required.

Polyester, being the single major polluting sector, requires maximum attention in this context, in terms of research as well as adaptation policy. An individual analysis of the dyes for each sector-shade pair shows that natural dyes cause significantly lower emissions in all cases except a very few, for example, polyester, which is due to not-so-developed technology. The fact that the ranking of the dyes, after the inclusion of the criteria of cost and fastness, has changed in some cases and unchanged in quite a few reiterates the overall feasibility of natural dyes. Other than the natural dye sahara, most other natural dyes have been able to retain their ranks even after including these criteria.

A well-designed policy should be framed for the promotion of natural dyes in particular and reduction of emissions in the textile dyeing sector in general. This should include features that would encourage the research in the application of natural dyes on the synthetic fibres and improvement of the colouring properties of the natural dyes. A meticulous action plan comprising of rates of adaptation to the natural dyes should be framed. A set of rules and regulations should be designed for the dyeing factories throughout the country for the adaptation of natural dyes. The natural dyes, which are found to be outstanding based on different criteria, as analysed in this study, should be chosen and promoted commercially.

If, as found in this study, the natural dyes are not feasible for certain applications, then synthetic dyes that are less polluting or are better than others as per the ranking should be promoted in such cases. Techno-economic feasibility analysis of natural dyes would prove useful in tracking the practical issues involved in the adaptation of natural dyes and hence should be encouraged. The standards of emissions in the effluents should be fixed in a scientifically reasonable manner and implemented strictly. One way of promoting natural dyes could be a systematic setting of the emission standards in such a way that the adaptation of natural dyes would be mandatory to achieve such standards.

In short, we come up with the following policy conclusions:

1. Promotion of natural dyes can reduce pollution due to the textile industry substantially. The current policy dilemma due to the failure of Effluent Treatment Plants (ETPs) and collective ETPs could be addressed well by this approach.
2. Given that natural dyes are not so compatible with synthetic fibres, policymakers could incentivize and promote research efforts in this area.
3. Stringent implementation of standards in water quality of rivers near the textile processing plants could be in effect a policy of promoting natural dyes.
4. Techno-economic feasibility studies indicate that natural dyes are both less polluting and cost-effective. Such analysis needs to be done on case-by-case basis.
5. If natural dyes are just not feasible, for example, due to the lack of fastness, policies should have some flexibility but promote less polluting synthetic dyes.

Several extensions of this study could be undertaken in the future. An AHP analysis could be done by including other criteria such as quality of the colour, durability and health hazards and collecting the opinion of experts from different areas. A cost-benefit analysis of natural dyes could be done by including various factors such as savings in ETP costs, cost-differences of dyes and environmental costs. Environmental costs of natural dyes, due to over-intensification of the land for their cultivation, could be analysed.

Organic dyes, which are currently under development and are claimed to be eco-friendly and better in quality and fastness than natural dyes, could be compared with natural dyes in terms of costs, emissions, etc. Inclusion of a measure of toxicity of the effluent, such as LC_{50}, could be done so that the toxicity of some natural dyes would also be internalized in the analysis. The extension of this analysis to some more sectors, dyes and scenarios based on different assumptions on the production could be done.

References

Agarwal KK (2003) Problems and prospects of using natural dyes in an industrial Enterprise. Colourage 50(6):37–40

Arasaratnam S (1996) Cloth and commerce: textiles in colonial India, Chapter

Auslander DE, Goldberg M, Hill JA, Weiss AL (1977) Naturally occurring colorants: A stability evaluation. Drug & Cosmetic Industry 121(5) 36, 38, 40, 105, 114

Bahl D, Gupta KC (1988) Development of dyeing process of silk with natural dye-cutch. Colourage 35(22):22–24

Bechtold T, Turcanu A, Ganglberger E, Geissler S (2003) Natural dyes in modern textile dye-houses — how to combine experiences of two centuries to meet the demands of the future? J Clean Prod 11:499–509

Chapman S (1972) The cotton industry in the industrial revolution. Macmillan Press Limited, London

Chaudhuri K (1996) Cloth and commerce: textiles in colonial India, chapter the structure of Indian textile industry in the seventeenth and eighteenth centuries, pages 33–84. Sage Publications, New Delhi

Das DK (2003) Quantifying trade barriers. Working Paper No: 105, Indian council for research on International Economic Relations, New Delhi

Dheeraj, Talreja, Priyanka, Talreja, Monika, Mathur (2003) Eco-friendliness of natural dyes. Colourage 50(7):35–36. 38,40–42,44

Duff DG, Sinclair RS, Stirling D (1977a) The fastness to washing of some natural dyestuffs on wool. Stud Conserv 22(4):170–176

Duff DG, Roy S, Stirling D (1977b) Light-induced color changes of natural dyes. Stud Conserv 22(4):161–169

Furry MS (1945) Some natural dyes give long life to cotton fabric. Rayon Tex Mon 26:603–605

Gillion KL (1968) Ahmedabad: a study in Indian urban history. University of California Press, Berkeley and Los Angeles

Glover B (1995) Are natural colorants good for your health? Are synthetic ones better? Text Chem Color 27(4):17–20

GoI (1990) The Textile Industry in the 1990s: restructuring with a human face report of the committee to review the progress of implementation of textile policy of june 1985. Number CMA Monograph 159. Government of India, New Delhi

Gokhale B (1979) Surat in the seventeenth century: a study in the urban history in pre-modern India, Number 28 in Scandinavian Institute of Asian Studies. Popular Prakashan, Bombay

Gokhale CS, Katti V (1995) Globalising Indian textiles: threats and opportunities. Tecoya Disseminators, Bombay

© The Author(s) 2018

B.N. Gopalakrishnan, *Economic and Environmental Policy Issues in Indian Textile and Apparel Industries*, SpringerBriefs in Environmental Science, DOI 10.1007/978-3-319-62344-3

Goswami O (1985) Indian textile industry, 1970-1984: an analysis of demand and supply. Econ Polit Wkly 20(38):1603–1614

Goswami O (1990) Sickness and growth of India's textile industry: analysis and policy options. Econ Polit Wkly 25(44):2429–2439

Govil K (1950) Cotton industry of India: Prospect and retrospect. Hind Kitabs Limited, Bombay

Guha S (1996) Cloth and commerce: textiles in colonial India, chapter the hand- loom industry of Central India:1825–1950, pages 218–241. Sage Publications, New Delhi

Gulati AN (1949) Natural Indian dyes and the art of their application. Indian Tex J 60:223–229

Haynes D (1996) Cloth and commerce: textiles in colonial India, chapter the dynamics of conti- nuity in Indian domestic industry: J ari manufacture in Surat, 1900–47, pages 299–325. Sage Publications, New Delhi

Holme IA (2003) Recent developments in colorants for textile applications. Colourage Annual 50(4):81–106

Hossain H (1988) The company weavers of Bengal. Oxford University Press, Delhi Hossain H (1996). Cloth and commerce: textiles in Colonial India, chapter The Alienation of weavers: the impact of the conflict between the revenue and commercial interests of the East India Company, 1750–1800, p 115–141. Sage Publications, New Delhi

Kharbade BV, Agrawal OP (1988) Analysis of natural dyes in Indian historic textiles. Stud Conserv 33(1):1–8

Kirtikar DB (1947) Indigenous. Indian dyes Indian Tex J 58:159–160,137

Kumar N (1995) The artisans of Banares: popular and cultural identity (1880–1986). Orient Longman, New Delhi

Majumdar R, Raychaudhuri H, Datta K (1963) An advanced history of India. Macmillan and Company Limited, London

Ministry of Textiles (2007) Review note on growth and investment in textiles during 2006–07. Economic Division, Ministry of Textiles, Government of India, New Delhi

Misra S (1993) India's textile sector: a policy analysis. Sage Publishers, New Delhi

Mitra DB (1978) The cotton weavers of Bengal. Firma KLM private limited, Calcutta

Moses JJ, Ravi N (2003) Application of grape skin powder extract on protein textile fabrics. Man-Made Tex India 46(8):295–300

Mukund K (1999) The trading world of the Tamil merchant: evolution of merchant capitalism in the Coromandel. Orient Longman, Hyderabad

Mukund K, Sundari BS (2001) Traditional industry in the new market econ- omy: the cotton han-dlooms of Andhra Pradesh. Sage Publications, New Delhi

Murty GVSN, Sukumari TR (1991) Demand for textiles in India. Econ Polit Wkly 26(21):M61–M67

Naik G, Babu K (1993) Demand and supply prospects for high quality raw silk, Number CMA monograph 159. Oxford and IBH Publishing Company, New Delhi

Narayanan GB (2005d) Exchange rate, productivity and exports: the case of Indian textile sector. J Indian Sch Polit Econ 17(4):657–678

Narayanan GB (2007a) Some economic issues in Indian Textile Sector. Unpublished PhD Thesis, Indira Gandhi Institute of Development Research, Mumbai, India

Narayanan GB (2007b) Economic studies of indigenous traditional knowledge, chapter pollution control by the natural dyes, pages 251–82. Academic Foundation in association with Indian Economic Association Trust for Research Development, New Delhi

Narayanan GB (2008b) Unveiling protectionism: regional responses to remaining barriers in the textiles and clothing trade, chapter Indian textile and apparel sector: an analysis of aspects related to domestic supply and demand, pages 129–156. In: Trade and investment division. Asia and Pacific, Bangkok, United Nations Economic and Social Commission for

Niranjana S, Vinayan S (2001) Report on growth and prospects of the handloom industry: study commissioned by the Planning Commission. Dastkar Andhra

Nousiainen P (1997) Modern textile dyeing takes note of the environment. Kemia - Kemi 24(5):376–380

NSSO (1989) NSS 40th round (July 1984–June 1985): tables with notes on survey of unorganised manufacture: non-directory establishments and own account enterprises, part I, part II (volume

1 and 2), Number 363/1. National Sample Survey Organisation, Department of Statistics, Government of India, New Delhi

NSSO (1994) NSS 45th round (July 1989–June 1990): tables with notes on survey of unorganised manufacture: non-directory establishments and own account enterprises, part-I (all-India), Number 396/2. National Sample Survey Organisation, Department of Statistics, Government of India, New Delhi

NSSO (1998) NSS 51st round (July 1994–June 1995): assets and borrowings of the unorganised manufacturing sector in India, Number 435. National Sample Survey Organisation, Department of Statistics, Government of India, New Delhi

NSSO (2002) NSS 56th round (July 2000 June 2001): unorganised manufacturing sector in India: characteristics of enterprises, Number 477. National Sample Survey Organisation, Department of Statistics, Government of India, New Delhi

NSSO (2005) NSS 60th round: household consumer expenditure in India, Number 505. National Sample Survey Organisation, Department of Statistics, Government of India, New Delhi

Parham RJ (1994) Elimination of production and environmental problems in cellulose dyeing: neutral dyeing reactives, bifunctional reactives, and natural dyes, Book of Papers – International Conference & Exhibition, AATCC. pp 398–403

Patra AK, Sareen A, Vohra D (2002) Performance studies of some natural dyes on cotton. Man-Made Textiles in India 45(8):319–323

Paul R, Jayesh M, Naik SR (1996) Natural dyes: classification, extraction and fastness properties. Textile Dyer & Printer 29(22):16–24

Pearson M (1976) Merchants and rulers in Gujarat: the response to portugese in the sixteenth century. Munshiram Manoharlal Publishers Pvt. Ltd.

Pophalia GR, Kaula SN, Mathur S (2003) Influence of hydraulic chock loads and TDS on the performance of large-scale CETPs treating textile effluents in India. Water Res 37:353–361

Ray P (1986) The cultural heritage of India, volume 6, chapter chemistry in ancient and medieval India, pages 136–151, 2nd edn. Bharatiya Vidya Bhavan, New Delhi

Ray I (2005) The silk industry in Bengal during colonial rule. Indian Econ Soc Hist Rev 42(3):339–376

Roberts R (1996) Cloth and commerce: textiles in colonial India, chapter West Africa and the Pondicherry textile industry, pages 142–174. Sage Publications, New Delhi

Roy T (1996a) Cloth and commerce: textiles in colonial India. Sage Publications, New Delhi

Roy T (1996b) Cloth and commerce: textiles in colonial India, chapter introduction, pages 11–32. Sage Publications, New Delhi

Rubeena S, Sharma RC, Pandey GP (2002) Biochemical analysis of natural dyes: an empirical study. Res J Chem Environ 6(4):67–72

Sahai NP (2005) Artisans, the state and the politics of wajabi in the eighteenth century Jaipur. Indian Econ Soc Hist Rev 42(1):41–68

Sastry DU (1984) The cotton mill sector in India. Oxford University Press, New Delhi

Schendel W v (1995) Reviving a rural industry: silk producers and officials in India and Bangladesh from 1880s to 1980s. Sage Publications, New Delhi

Siebert H, Eichberger J, Gronych R, Pethig R (1980) Trade and environment: a theoretical enquiry. Elsevier/North Holland Press, Amsterdam

Specker K (1996) Cloth and commerce: textiles in colonial India, chapter madras handlooms in the nineteenth century, pages 175–217. Sage Publications, New Delhi

Swarnalatha P (2005) The world of the weaver in northern Coromandel c.1750- c.1850. Orient Longman, Hyderabad

Teli MD, Paul R, Pardeshi PD (2000) Natural dyes: classification, chemistry and extraction methods part - I: chemical classes, extraction methods and future prospects. Colourage 47(12):43–48

Uchikawa S (1998) India's textile industry:state policy, liberalisation and growth. Manohar Publishers, New Delhi

Welfare: The role of technology (2006) Applied economics letters, 13(1): 63–66

Yanagisawa H (1996) Cloth and commerce: textiles in colonial India, chapter the handloom industry and its market structure: the case of the madras presidency in the first half of the twentieth century, pages 242–273. Sage Publications, New Delhi

Printed in the United States
By Bookmasters